RUSSIAN RENEWABLE ENERGY

Russian Renewable Energy
The Potential for International Cooperation

INDRA ØVERLAND and HEIDI KJÆRNET
*Norwegian Institute of International Affairs (NUPI)
and University of Tromsø, Norway*

LONDON AND NEW YORK

First published 2009 by Ashgate Publishing

Published 2016 by Routledge
2 Park Square, Milton Park, Abingdon, Oxon OX14 4RN
711 Third Avenue, New York, NY 10017, USA

Routledge is an imprint of the Taylor & Francis Group, an informa business

British Library Cataloguing in Publication Data
Øverland, Indra.
  Russian renewable energy : the potential for international
  cooperation.
  1. Renewable energy sources--Russia (Federation) 2. Power
  resources--Research--Russia (Federation) 3. Power
  resources--Research--International cooperation.
  I. Title II. Kjærnet, Heidi.
  333.7'9415'0947-dc22

Library of Congress Cataloging-in-Publication Data
Øverland, Indra
  Russian renewable energy : the potential for international cooperation / by Indra Øverland and Heidi Kjærnet.
    p. cm.
    Includes index.
    ISBN 978-0-7546-7972-1 (hardback)
    1. Power resources--Government policy--Russia (Federation). 2. Renewable energy sources--Russia (Federation) 3. Russia (Federation)--Foreign economic relations. I. Kjærnet, Heidi II. Title.

  HD9502.A4O82 2009
  333.79'40947--dc22

                                                                        2009031153

ISBN 9780754679721 (hbk)

Contents

List of Figures

List of Tables

Notes on Contributors

Indra Øverland is the Head of the Energy Programme at the Norwegian Institute of International Affairs (NUPI) and Associate Professor at the University of Tromso. His main research interest is post-Soviet energy issues. He holds a PhD from the University of Cambridge and is the co-editor of numerous scholarly articles, as well as a book to be published by Routledge in September 2009 (*Caspian Energy Politics: Azerbaijan, Kazakhstan and Turkmenistan*).

Heidi Kjærnet is a Research Fellow under the Energy Programme at NUPI. Her current research focuses on energy policy and government–business relations in Azerbaijan, Kazakhstan and Russia. She has served as project manager for the environmental organization Nature and Youth in North-Western Russia, and is the co-editor of *Caspian Energy Politics: Azerbaijan, Kazakhstan and Turkmenistan*, to be published by Routledge in September 2009.

Acknowledgments

This book is based on a comprehensive study of renewable energy in Russia carried out between 1 November 2007 and 15 June 2009. The study was sponsored mainly by the organization Nordic Energy Research, but also drew on the resources of the RussCasp project. Nordic Energy Research, an institution under the umbrella of the Nordic Council of Ministers, supports joint energy research between the Nordic countries. RussCasp is financed by the Petrosam Programme of the Research Council of Norway. RussCasp is carried out by the Fridtjof Nansen Institute, the Norwegian Institute for International Affairs and Econ Pöyry as consortium partners, and also includes other institutions and researchers as participants. We would also like to thank Grant Dansie who co-authored Chapters 3, 5 and 7 as well as Nina Kristine Madsen who wrote Chapter 6.

We are extremely grateful to Marc Lanteigne, Diana Golikova, Nazar Mamedov, Farkhod Aminjonov and Susan Høivik for their help with language editing, formatting, layout and helping us to sort out all the stylistic details of the book.

Chapter 1
Introduction

Renewable energy has emerged as a primary tool in the global strategic race towards a low-carbon economy. Countries that are successful in this race will gain economic strength, while making a contribution to climate policy that may raise their international political standing and reduce their dependency on imported energy. The ability to forge fruitful partnerships across borders will be a decisive factor. In this book, we examine whether and in what ways Russia might be suitable for such partnership. We seek to provide an overview of the Russian renewable energy landscape that may:

- help foreign governments and research funds to formulate policies to promote international cooperation with Russia on renewable energy,
- help international researchers to identify which Russian research institutions, sectors and locations may be worth targeting for collaborative ventures,
- provide students, researchers and practitioners with insights into the situation in Russia in the field of renewable energy as well as its future potential,
- hold up a mirror, making an outside view available to Russian researchers and thereby help them develop their own renewable energy sector,
- shed light on Russia's strengths in renewable energy, as well as the obstacles to renewable energy research and investment in the country.

The book highlights the vast potential that Russia has in renewable energy, as well as obstacles to renewable energy research and investment in the country. Relatively little information is available on Russia and renewable energy, so this volume should be of interest to readers generally involved in global renewable energy issues, as well as those more specifically interested in Russian politics and economics, or in the Eurasian energy balance.

In addition to providing a detailed picture of the actors and framework conditions in Russia's renewable energy sector, we examine cooperation between the European Union (EU) and the Russian Federation, and use the cooperation between the Nordic countries and Russia as an informative case study. The EU is Russia's largest trade partner, and within the EU the Nordic countries have spearheaded relations with Russia. There is much to learn from the experience of the Nordic countries, with their proximity to Russia, the intensity of their broader economic cooperation with Russia and their advanced technologies for renewable energy.

While considerable academic work has been done on Russia's petroleum sector, renewable energy in Russia has remained less studied. Nonetheless renewable energy in Russia is a timely topic, as we explain in the following section. In the final part of this introductory chapter we go on to outline the topics covered in this book and its structure.

Why renewable energy in Russia?

Why pay attention to renewable energy in Russia – a country so richly endowed with fossil fuels? After all, it holds the world's largest reserves of natural gas, the second largest coal reserves, eighth largest oil reserves, and is the world's largest exporter of natural gas, the second largest oil exporter, one of the main nuclear powers and the world's largest energy exporter.[1] Russia's position as an energy superpower is based on its natural abundance of fossil and nuclear fuels – not 'wishy-washy' renewable energy.

In fact, renewable energy is highly relevant for Russia, for several reasons. First of all, Russia can benefit economically from giving greater priority to renewable energy sources, since this will improve its opportunities for energy exports by decreasing the domestic use of fossil fuels. Secondly, Russia can make the transition to renewable energy usage more cost-efficient and cheaper by using mechanisms in the global climate regime that promote increased production of renewable energy. Thirdly, whereas fossil fuels are exhaustible, finite resources, renewable energy sources are not. This means that developing renewable energy sources will be necessary sooner or later. Being proactive in this development will undoubtedly be an advantage. Fourthly, Russia's vast size means that renewable energy solutions are the most economically viable option in certain isolated areas such as the Northwest. And finally, Russia enjoys several competitive advantages linked to its natural resource base and its strong tradition of research in the natural and technological sciences.

As Russia's position as an emerging energy superpower is based primarily on its natural abundance of fossil and nuclear fuels, one might imagine that the country's increasing international leverage is exclusively related to its conventional energy sources, and has little to do with the 'softer' renewable energy forms. As we shall see below, however, the two are intertwined.

Russia and the international climate regime

The Kyoto Protocol under the International Framework Convention on Climate Change, aimed at reducing the greenhouse gases that contribute to climate change,

---

1   Energy Information Administration. 'Russia Energy Profile', May 2008. http://tonto.eia.doe.gov/country/country_energy_data.cfm?fips=RS [accessed 9 June 2008].

entered into force in 2005. In 2008 the global emissions trade system became effective. The first period of the Protocol runs until 2012, and the second period is currently under negotiation. Russia ratified the Kyoto Protocol in 2003. Since it accounted for 16.4 per cent of 1990 carbon emissions, and since Annex B parties accounting for together 55 per cent of 1990 carbon emissions had to ratify, in order for the Kyoto Protocol to enter into force, Russia's ratification tipped the scales and triggered activation of the Protocol. Paradoxically enough – given the low priority of climate issues on the Russian policy agenda, and the scepticism of Russian scientists to the linkage between greenhouse gas emissions and climate change – it was Russia's ratification that brought the Kyoto Protocol into force.[2] The Russian debate on climate politics has emphasized the potential for wealth redistribution that lies in the Kyoto mechanisms (for example, through technology transfers), and the Russian negotiators in the ratification process stressed side-benefits like potential WTO accession.[3]

So far, Russia has not had any problems in achieving its Kyoto target level. The target level of greenhouse gas emissions set for Russia in the first period of the Kyoto Protocol (2008–2012) is the equivalent of its 1990 emissions. Greenhouse gas emissions were at the highest level in Soviet history in 1990, which made this year a logical choice as Russia's target level.[4] By 1995, emission levels of greenhouse gases had fallen by approximately 40 per cent.[5] However, this reduction was not a result of structural changes or technological improvements in the industry that led to actual reduction in emissions from ongoing production: it was more the result of factories being shut down due to the collapse of the Soviet economy in 1991.

The difference between a country's business-as-usual emissions and its official emissions ceiling is referred to as 'hot air'. Under the emissions trade system, this can be sold as a right to emit equal amounts of greenhouse gases elsewhere. The possibility of profiting from the sale of 'hot air' emission quotas without having to reduce emissions further appears to be an important reason why the Russian government chose to ratify the Kyoto Protocol in October 2004 – only six months after then-President Vladimir Putin's ultra-liberalist economic adviser

---

2   Moe, Arild. 'The Kyoto Mechanisms and Russian Climate Politics'. Presentation at the conference *Renewable Energy in Russia: How Can Nordic and Russian Actors Work Together?*, Oslo, Norway, 8 May 2008.

3   Moe 2008.

4   1990 is the base year for most countries in the Kyoto Protocol. Being an economy in transition, under Article 3.5 of the Protocol, Russia could however have selected a different base year, either 1990 or any earlier year. See Golub, Alexander and Elena Strukova. 'Russia and the GHG Market'. *Climatic Change*, no. 63 (2004): 223–43, p. 225.

5   Oldfield, Jonathan D. *Russian Nature: Exploring the Environmental Consequences of Societal Change.* Aldershot: Ashgate, 2005, p. 56.

Andrey Illarionov had denounced the work of the UN International Climate Panel as following a totalitarian ideology, and had claimed that the link between climate change and carbon emissions was not scientifically proven.[6] There were also hopes on the Russian side that the EU countries would support Russian WTO membership in return for ratification of the Protocol.[7]

For the first five-year period of the Kyoto Protocol, relying on the 'hot air' quotas will probably be enough for Russia to achieve its Kyoto target level.[8] Despite its seeming unwillingness to accept the global climate change paradigm, and depending on how the market for emissions trading develops, the Russian leadership might actually find itself benefiting economically from implementing the climate regime, due to the 'hot air' quotas. However, post-*perestroika* Russia is not a country where environmental issues enjoy priority; moreover, being highly dependent on oil and gas revenues, it is likely to consider the effect of the availability of 'hot air' on oil and gas prices. If 'hot air' emission permits were to become less readily available, that would force greater reductions of energy consumption in other countries, which in turn could cause a reduction in the international value of fossil fuels. That would mean a loss of Russian export revenues from oil and gas exports. There is likely to be a trade-off between the revenues from permits sales, and revenues from oil and gas. The outcome will depend on petroleum prices, and price developments in the emissions trade market.

The decline in Russia's greenhouse gas emissions in the early 1990s was steep, but not as steep as the decline in its GDP. That meant that the Russian economy became more carbon-intensive, among other things because the natural resource sector contracted less rapidly than other sectors of the economy.[9] The Russian economy is highly energy-intensive, and the country's energy needs are met primarily by fossil fuels. In 2005, natural gas accounted for 54 per cent of primary energy consumption in Russia, while oil and coal accounted for 19 per cent and 16 per cent respectively. Altogether 89 per cent of the country's energy consumption is met by sources that cause greenhouse gas emissions. The remainder is accounted for by nuclear energy (5 per cent) and hydropower (6 per cent – this estimate includes big hydro).[10] Nevertheless, since the 1998 crash and devaluation of the rouble, Russia has experienced a rapid economic recovery, and in 2007 GDP reached the 1990 level (in absolute terms) – although its greenhouse gas emissions did not.

---

6 *Russian Courier.* 'Putin's Aid: Kyoto Protocol is Totalitarian', 19 May 2004. http://www.gateway2russia.com/st/art_236288.php [accessed 5 May 2008].

7 Oldfield 2005, p. 137.

8 Golub and Strukova 2004.

9 Golub and Strukova 2004, p. 224.

10 BP statistics quoted in *Russian Analytical Digest.* 'Russia's Energy Policy', no. 18 (3 April 2007): 1–17.

Emission target levels are currently being negotiated for the second period of the Kyoto Protocol, due to start in 2012. It seems likely that the increasing acuteness of the climate issue will make it necessary for all signatories to take on tougher commitments, which in turn will necessitate structural changes in energy systems. Russia, which obviously had an easy target in the first Kyoto period, and which has now recovered from the economic collapse of 1998, should be prepared to meet demands to accept heavier commitments at this crossroads. The main source of greenhouse gas emissions in Russia is the energy sector. Developing renewable energy sources and energy efficiency measures could therefore contribute greatly to the structural changes needed for the country to assume tougher commitments without slowing its economic growth.

However, Russia seems unwilling to give environmental issues priority over economic development – a stance that is in no way unique and was mirrored for example in the US position under the George W. Bush administration, and the former Australian position under Prime Minister John Howard.

Increased export opportunities

If the climate regime is not top priority for Russian decision-makers, economic development certainly is. The possibilities offered by renewable energy and energy efficiency in terms of increasing energy exports should therefore catch their interest. The importance of efficient and reliable electricity services for economic development within the country is also clear. Russia is pursuing a strategy of high economic growth, aiming to double its GDP in ten years. Efficient and reliable electricity markets will be critical to the success of this policy.[11]

By increasing its production of renewable energy and realizing its vast potential in energy efficiency, Russia could make a significant contribution towards improving the Eurasian gas balance, and ultimately strengthen its own importance as an energy provider for European gas markets.[12] Currently, there is increasing concern in the EU about declining production in the Nadym Pur Taz area of northwestern Siberia, underinvestment in the Russian gas sector and the slow development of new fields.[13] There is also the risk that mutual insecurities

---

11   IEA. *Russian Electricity Reform: Emerging Challenges and Opportunities.* Paris: IEA, 2005. http://www.iea.org/Textbase/publications/free_new_Desc.asp?PUBS_ID=1473 [accessed 29 February 2008].

12   The RussCasp research project run by FNI, NUPI and ECON is exploring this issue by conducting research on energy efficiency as a possible key to the Eurasian gas balance towards 2020.

13   Perović, Jeronim. 'Russia's Energy Policy: Should Europe Worry?'. *Russian Analytical Digest*, no. 18 (3 April 2007): 2–7.

between the EU and Russia could contribute to a race to diversify purchases and sales away from each other, even though existing mutual dependencies make this outcome undesirable to both sides.[14] By decreasing its domestic gas consumption, to which the introduction of renewables and greater energy efficiency could contribute, Russia could increase its energy exports.

Russia is one of the world's least efficient countries in terms of the amount of energy it uses. According to the Energy Strategy announced by the Russian Ministry of Energy in 2003, Russia expends 3.1 times the amount of energy that the EU uses to produce one unit of GDP.[15] In the Energy Strategy, the Ministry indicates that Russia could save half of the energy it currently uses. Economizing energy on this scale would allow more gas to be sold abroad, in turn contributing to Russia's economic growth as well as its importance for Europe. According to the Russian Energy Strategy towards 2020, the export of energy resources can grow by 45 to 64 per cent by 2020, which would strengthen the country's economic position and geopolitical influence, while taking into account the interests of the next generation of Russians.[16] This not only shows that decision-makers are aware of the potential energy efficiency gains, it also brings out the link between the country's economic potential and the expected gains in geopolitical leverage.

Adjusting to future energy systems

Russia has the advantage of vast geographic size and variation in climate and terrain, giving it the potential to develop virtually any kind of renewable energy.[17] This comparative advantage in renewables, along with the extreme inefficiency in current energy use, provides Russia with considerable potential for contributing to greener energy use on a global scale.

Russia's Federal Law on Energy Saving defines renewable energy sources as 'solar energy, wind, earth thermo energy, natural hydro movement and nature heat production' – thereby excluding traditional large-scale hydroelectricity.[18] Estimates of the potential of renewable energy can be variously calculated. For example, estimates of available resources can tell us the energetic equivalent of

---

14   Perović 2007.

15   Ministry of Energy of the Russian Federation. *Summary of the Energy Strategy of Russia for the Period of up to 2020*. Moscow: Ministry of Energy of the Russian Federation, 2003.

16   Ministry of Energy of the Russian Federation 2003.

17   IEA. *Renewables in Russia: From Opportunity to Reality*. Paris: IEA, 2003. http://www.iea.org/textbase/nppdf/free/2000/renewrus_2003.pdf [accessed 29 February 2008].

18   Brown, Anna. 'Russian Renewable Energy Market: Design and Implementation of National Policy'. *Russian/CIS Energy and Mining Law Journal*, vol. 3, no. 6 (2005): 33–9.

Table 1.1 **Russia's renewable energy resources, PEEREA estimates, Mtoe/ year[19]**

Resource	Technical potential	Economic potential
Small hydro	88	49
Geothermal	–	80
Wind energy	1400	8
Biomass energy	37	5
Solar energy	1610	2
Low-grade heat	136	37
TOTAL	3271	181

the total amount of renewable energy available for extraction. Taking into account the technological limitations as well as the social and ecological factors yields what is referred to as the 'technical potential'. What is really decisive – the *economic* potential – is that part of the technical potential which is economically justified when one takes into account the costs of fossil fuels, heat and electricity, equipment, materials, transportation and wages.[20] This means that the economic potential of renewable energy sources will increase with the rise in fossil fuel prices.[21]

Russia, as the biggest country in the world, has natural preconditions that grant it a competitive advantage in the development of renewable energy. With its vast and varied territory, the country has so many different types of nature that there are few energy sources that do not exist there. However, it is difficult to ascertain the exact potential of the various resources, and estimates diverge. The Energy Charter Protocol on Energy Efficiency and Related Environmental Aspects (PEEREA) from 2007 provides one set of estimates, as shown in Table 1.1.

Pavel Bezrukikh, Deputy Director of the Russian State Institute of Energy Strategy, has made another widely quoted estimate of the renewable energy potential in Russia, which is presented in Table 1.2. He found that the economic potential for development of renewable energy sources in Russia could cover 35 per cent of the country's total primary energy supply (TPES). When we compare this with today's situation, where renewable energy sources account for less than 1 per cent of Russia's energy, two points stand out: renewable energy sources

---

19 Source of data used to compile table: EC. *The Energy Charter Protocol on Energy Efficiency and Related Environmental Aspects: Regular review of Energy Efficiency Policies.* Brussels: EC, 2007a.

20 IEA 2003.

21 This is highly relevant for Russia, which has heavily subsidized natural gas for domestic consumption.

Table 1.2   Russia's estimated renewable energy potential[22]

Energy source	Potential (Mtoe)	
	Technical	Economic
Wind	1551.2	7.7
Small hydro	88.2	49
Solar	6786.5	2.1
Biomass	90.3	48.3
Geothermal	8308.3	79.8
Low-potential heat	135.8	37.1
Total	16981	228

are underdeveloped; and renewable energy sources can contribute tremendously to the energy balance, export potential, emissions trade potential and economic development of the Russian Federation.

The International Energy Agency survey from 2003 found that wind energy can be exploited in many parts of Russia – including Arkhangelsk, Astrakhan, Volgograd, Kaliningrad, Magadan, Novosibirsk, Perm, Rostov, the Tyumen regions, Krasnodar, Khabarovsk, maritime territories, Dagestan, Kalmykia and Karelia. According to the same source, the country's solar energy potential is greatest and can be exploited in the southwest (the North Caucasus, the Black Sea and Caspian Sea regions) as well as in Southern Siberia and the Far East. The IEA survey found geothermal resources viable for production in several places, notably on the Kamchatka Peninsula and Kuril Islands, but also in the North Caucasus as well as certain locations in Central Russia, Western Siberia, Lake Baikal, Krasnoyarsk Krai, Chukotka and Sakhalin.[23] In addition, Russia is the world's largest producer of biomass, which means that it also has a tremendous potential for bioenergy.[24]

Contrasting Russia's potential for renewable energy with its installed capacity reveals considerable scope for expansion. (See Table 1.3.)

---

22   Source of data used to compile table: Pavel Bezrukikh, quoted in Merle-Béral, Elena. 'The Wider Perspective: Russia's Energy Scene'. Presentation at the conference *Renewable Energy in Russia: How Can Nordic and Russian Actors Work Together?*, Oslo, Norway, 8 May 2008.

23   Merle-Béral, Elena. 'Russia Renewable Energy Markets and Policies: Key Trends'. Presentation at *Global Best Practice in Renewable Energy Policy Making*, expert meeting, Paris, France, 29 June 2007. http://www.iea.org/Textbase/work/2007/ bestpractice/Merle_Beral.pdf [accessed 23 July 2008].

24   Brown 2005, p. 35.

Table 1.3   Installed renewable energy capacity[25]

Type	Installed capacity, MW
Geothermal energy	73
Small hydro	1,000
Large hydro	46,000
Biomass	1,270
Wind	14
Tidal	0.4 (not currently operative)

In addition there are hydro-projects planned (capacity: 5GW by 2010), as well as wind projects in Kalmykia (23 MW), Kaliningrad (50 MW) and Primorsk (30 MW). Two tidal projects are planned, one at Mezenskaya (19 GW) and another at Tugurskaya (9MW).[26] There may also be other projects in the pipeline, but the lack of comprehensive data makes it difficult to present a complete overview.

Energy efficiency

Russia's industrial sector is highly energy-intensive. It is, however, not only in industry that energy use is inefficient. It is estimated that 30 to 40 per cent of all Russian energy is lost in production, transport, transmission or inefficient consumption.[27] This is why Russia is sometimes referred to as 'the Saudi Arabia of energy efficiency'.[28] The energy currently being wasted could in fact become the country's largest energy source. Industry and residential costumers alike are wasteful in dealing with energy. An important reason for this is how energy has been priced. Prices have not reflected production costs: ever since the Soviet period, access to inexpensive energy has been granted to industry and private consumers alike. In a market economy, this is not sustainable, and energy prices are set to increase.

A case study from Kirovsk and Apatity illustrates the potential for saving energy at all levels, for both industrial and residential energy consumption, throughout Russia.[29] The mining industry in these two cities has an energy consumption four

---

25   Source of data used to compile table: IEA 2003.

26   IEA 2003.

27   Aron, Leon. 'Privatizing Russia's Electricity', *Russian Expert Review*, no. 4 (2003): 9–17, p. 11.

28   Merle-Béral 2008.

29   Keikkala, Gudrun, Andrey Kask, Jan Dahl, Vladimir Malyshev and Viktor Kotomkin. 'Estimation of the Potential for Reduced Greenhouse Gas Emission in North-East *[sic]* Russia: A Comparison of Energy Use in Mining, Mineral Processing and Residential Heating in Kiruna and Kirovsk-Apatity'. *Energy Policy*, vol. 35, no. 3 (2007): 1452–63.

times higher per tonne of raw ore and six to seven times higher per ton of product, compared to a comparable mining company in Kiruna in Sweden. With regard to residential consumers, the same case study indicates an energy efficiency potential of 30 to 35 per cent, and in distribution in the district heating system losses can be reduced by 30 per cent. Despite its consistent references to the Kola Peninsula as being in the 'Northeast' of Russia (!), that case study convincingly shows how great the energy-saving potential is at all levels of the Russian energy system. Most fixed infrastructure in the system was built up during the Soviet period, and the structural flaws as well as the inadequacies in maintenance are similar throughout the nation. This means that the potential for energy saving can be assumed to be equally great in other parts of Russia.

Research tradition

Finally, Russia has a clear advantage with regard to the development of renewable energy due to its long traditions of high-quality scientific research. Ever since the Soviet period, there has been a heavy emphasis on education and training within the natural sciences. The country also has a strong track record of scientific research specifically on renewable energy technologies. In the early 1930s, the USSR constructed the first utility-scale wind turbines in the world. The first Russian atlas of wind energy resources was published in 1935. Over 7000 small-scale hydropower stations were built in the late 1940s. Research on photovoltaic cells was advanced due to the space programme, and the first solar-powered satellite, Sputnik 3, went into orbit in 1958. The 5 MW Pauzhetskaya geothermal power station was completed in 1967, and a 450 kW tide power station was built in 1968.[30] Unfortunately, this tradition suffered in the late 1960s and 1970s, when Soviet central planners came to favour nuclear energy and fossil fuels. Further details of Russia's scientific and educational system are presented in Chapter 3.

As an example of how Russia's strong scientific tradition could contribute to developing renewable energy, it is tempting to note the case of Google. The world's leading online search engine was founded by the American Larry Page and Russian-born Sergei Brin. The latter is the son of a mathematician and an economist from the Soviet Union, and his background is very much rooted in Russia's strong academic traditions. Brin contributed some of the key innovations to Google's search technology. Google earned 16.5 billion USD in 2007, a 56 per cent increase from 2006, and is currently valued at 137.8 billion USD.[31]

---

30   IEA 2003.
31   Hagen, Guro Aardal. 'Neppe et must for Google'. *Dagens IT* (2 June 2008), http://www.dagensit.no/finans/article1415158.ece?jgo=c1_re_left_6&WT.svl=article_title [accessed 6 June 2008].

Russian renewables: A future object of investment?

We have listed three reasons why Russia should improve its efforts in energy efficiency and development of renewable energy sources. Admittedly, the reasons relating to environmental concerns are not at the forefront of Russia's current priorities. However, the possibility of increasing the country's energy export potential and its economic growth should attract the attention of Russian decision-makers, so the sector may well expand in the near future.

In the hunt for new markets and new competitive advantages, some of the largest companies in the world have moved into the Russian market over the past decade or so, as exemplified by the investments of major Nordic companies like IKEA, Telenor, Carlsberg and Fortum. In spite of a range of severe problems including the manipulation of the legal system as a bargaining tool in business negotiations, several of these Nordic majors have made some of their biggest profits in Russia. Might not renewable energy prove to be a suitable future object of investment and cooperation with Russia?

Conditions for renewable energy in Russia

As pointed out above, with its geographical size and great variation in climate and terrain, Russia has the potential to develop virtually any kind of renewable energy. There are nevertheless difficulties in ascertaining the exact potential of the various resources, and estimates diverge. Contrasting Russia's potential for renewable energy with installed capacity, one can see that there is considerable scope for expansion. The country's size, and the consequent proximity of the various parts of the Russian Federation to Western Europe, China and Japan, may also be an advantage in terms of global market reach.

We have also noted how Russia is among the world's least efficient countries in its energy consumption, with some 45 per cent of its primary energy consumption wasted due to energy inefficiency.[32] Both industrial and residential customers have been spendthrift in their energy use, largely due to underpricing by state-controlled utilities. This is an unsustainable legacy from the Soviet period, and energy prices are now set to increase. Russia's comparative advantage in renewables, along with the extreme inefficiency in current energy use, indicates considerable potential for contributing to greener energy use. In contrast to the Russian leadership's vacillation between lip service and indifference to renewable energy, improving energy efficiency has now been recognized as important in the country's official energy strategy.

---

32 World Bank in Russia. *Energy Efficiency in Russia: Untapped Reserves*. Moscow: World Bank, 2008.

Exploring the conditions for the development of renewable energy sources in Russia, this book features an analysis of the ongoing reform of the country's electricity sector. The reform redefines the rules of the game, liberalizes the sector through the introduction of competition in some segments and also involves privatization of the state-owned monopoly RAO UES. The reform isolates one part of the electricity sector (transmission lines: both high-voltage and low-voltage), to be kept as a natural monopoly, while competition is introduced in generation and retail. The government will continue to regulate prices.

Much is likely to depend on not just the design of the reform, but its implementation. Examples from countries that have had a more developed market economy, a better regulatory framework and more transparent business environments than Russia have clearly shown the pitfalls in this area. Russia might end up repeating some significant errors. The likelihood that the reform will create substantial levels of market power and even monopoly power under peak-load conditions in certain regions is a serious threat to the development of renewable energy, as it will create considerable barriers to market entry. This calls for establishing truly independent regulatory mechanisms, and highlights the necessity of creating incentives and building the right institutions to enable Russia to realize its potential in energy efficiency and renewable energy.

The major impediment to the development of renewable energy in Russia is, however, the continued existence of subsidies for domestic gas consumption. These subsidies remain a major issue in the negotiations for Russian WTO membership, as prices are not expected to reach the European market level before 2014/15, given current trends. This market distortion is an impediment to the profitability of other energy sources, justifying the charge that the Russian market is not open to fair competition between non-renewable and renewable energy sources today.

Markets for renewable energy in Russia

Research on renewable energy in Russia has tended to deal mainly with the natural resource base for renewable energy, or with technologies for specific forms of renewable energy. By contrast, this book focuses on a topic that has received much less attention: the market potential for renewable energy.

Several factors that make Russia a potentially strong market for renewable energy: the steady growth of the economy and of the purchasing power of the population since 1998; the large numbers of summer cottages or *dachas* (an estimated 22 million); and an estimated five million farms and ten million people that are not connected to the central grids. One should, however, also bear in mind the limitations on the emergence of markets for renewable energy in Russia. We have mentioned the subsidies for natural gas and electricity. Additionally, gathering reliable statistical data can be difficult in Russia, and this lack of data can be an obstacle to the development of renewable energy markets, as it forces investors and innovators to operate in uncharted territory.

Here we should note that there is one niche where renewable energy might be able to gain considerable ground, without having to compete with subsidized natural gas and nuclear power: the remote settlements in the northern parts of the country without access to central electricity and gas grids. Northern areas with harsh climatic conditions make up around 60 per cent of Russia's territory. Yet, with less than one person per km², it is prohibitively expensive to connect most of these areas to the central grids. Additionally, transporting coal, gas and other energy commodities is extremely costly. These settlements could emerge as one of the first realistic market niches for the profitable implementation of renewable energy in Russia, pending the growth of better framework conditions in the rest of the country. As such, they could function as a testing ground for renewable energy, preparing the country and local as well as foreign actors for future expansion in this sector.

International projects in this niche market can count as joint implementation (JI) projects under the Kyoto Protocol. This means they can benefit the climate accounts of foreign partner countries as well as reducing overall emissions.

The Russian scientific-educational system

The book contains an introduction to the Russian scientific-educational system in order to help actors interested in scientific and education collaboration with Russia to understand the country's long history of research and vast landscape of scientific and educational institutions. We also present an overview of organizations that fund research in Russia, which may be of use to international research funding organizations wishing to make joint calls with Russian funding institutions.

The economic chaos of the 1990s had severe consequences, and there has been a double brain-drain from Russian science: from research to other sectors of the Russian economy, and from Russia to other countries. The decline in funding and the lack of opportunities for young scientists are interrelated and equally negative developments. Recently the outlook has improved slightly, thanks to greater governmental focus on science and some of Russia's largest companies maturing, with prospects for heightened interest and investment in research. The research and development share of Russian GDP has increased, and the disbursement of state funding is now more timely and predictable than during the difficult 1990s.

The book contains a table listing some central Russian research funding institutions that could be relevant for international actors interested in co-financing renewable energy projects. We have also prepared a ranking of the leading Russian research and education institutions in renewable energy, to serve as a guide for those searching for Russian partners for EU or other international projects. The ranking offers insight into what may be the top institutions when it comes to renewable energy in Russia. All the institutions ranked are presented in a table with web addresses.

Innovation

The economic growth currently experienced by Russia is heavily dependent on high commodity prices. Focusing on research and innovation would help to foster new industries, increase productivity and diversify the economy. Russian innovation indicators remain disappointing, although the potential remains great, due to a firm science base and an educational system strong on science and technology.

Russian research and development is still primarily financed by the state. Innovation has remained low in the private sector, which has focused on imitation rather than research-based innovation. Poor communication between the public and private sectors has also affected the levels of commercialization. With most research and innovation being state-funded, researchers in the public sphere have generally little incentive to concern themselves with the commercial applications of their work. A second major constraint on commercialization is Russia's weak intellectual property rights framework. Educational backgrounds can also have an effect on commercialization, insofar as many educational programmes still do not fully prepare students for market-oriented work. International actors aiming to carry out commercialization in cooperation with Russian partners must therefore explicitly highlight both the benefits to be achieved from commercialization and the opportunities available.

Kyoto mechanisms

Due to its position as one of the most energy-intensive economies in the world, as well as one of the most inefficient and wasteful, Russia offers a prime opportunity for Joint Implementation (JI) projects under the Kyoto Protocol. This can enable EU countries to fulfil their commitments to reducing emissions under the Kyoto Protocol, while at the same time establishing a presence in the Russian renewable energy sector. Of the 163 JI projects currently underway in the world, 109 are located in Russia, the majority of them initiated by actors in the EU countries.

Learning from the Nordic experience

A survey of cooperation between the Nordic countries and Russia on renewable energy is included as a case study in this book. Cooperation between the Nordic countries and Russia is relevant to a broader audience, for several reasons: the Nordic countries are world leaders in renewable energy technology; the Nordic region, unlike the USA, Japan or most EU countries, shares a long land border with Russia; and Nordic companies have invested heavily in the Russian economy, more so on average than the companies of other OECD countries. For all these reasons, the richest experience of renewable-energy cooperation with Russia is to be found in the Nordic countries, and can offer important lessons for other actors.

A large number of pan-Nordic and pan-Baltic institutions have been created to foster regional collaboration involving Russia. However it appears that there is significant overlap between the goals and aims of the various multilateral organizations. They form a complex interwoven web, sharing focus areas and often working on and financing the same projects. The nature of this web and the shared priorities make it difficult to identify which actors have the best expertise in specific areas. A weakness of some Nordic funding is that it has been spread too thinly and between too many actors to serve as the main financing for serious research projects. This is especially the case when dealing with Russian actors, who will rarely be familiar with the Nordic institutions in the first place. This experience may be of particular relevance for EU organizations, which also often comprise multifarious institutional webs.

Nordic actors seeking cooperation with Russians on renewable energy have tended to focus on setting up generation capacity for renewable energy for the Russian electricity market. In fact, a more appropriate angle might be to search for high-level competence in areas of basic science that are central to the development of renewable energy technologies. Any manufacturing needed could take place in Russia or elsewhere, and the most realistic markets at present are outside Russia. Solar power and hydrogen technologies are two areas where Russian science has traditionally been strong and where researchers and companies could contribute significantly to international projects.

The Nordic experience highlights a range of challenges that actors wishing to collaborate with Russian scientists will need to grapple with. There is a significant degree of bureaucracy, and finding the right researchers can be difficult. Disagreements between foreign and Russian scientists about the anthropogenic causes of climate change may be a problem as well. More general communication and cultural differences also pose challenges.

Collaboration with the Nordic countries has also been limited due to a lack of co-financing from the Russian state. One reason could be the combined Nordic-centric and aid-oriented character of many of these projects, giving the Russian side little incentive to contribute. However, this also reflects a lack of interest in renewable energy on the part of Russia's federal government. Here it should also be noted that the Russian government is more than willing to allow Nordic actors to continue to undertake energy efficiency projects, as long as most of the funding keeps coming from the Nordic side. At the same time there is considerable potential for private partnerships and working with NGOs in Russia.

In addition to the general overview of European and Nordic cooperation with Russian actors on renewable energy, we have conducted a study of the subjective experiences of ten Nordic actors who have been involved in collaboration with Russian counterparts on renewable energy projects. We find that perceived opportunities are mostly related to Russia's natural resources, mainly hydro, wind and bioenergy. The challenges relating especially to cultural differences, corruption and bureaucracy are described at length, based on the experiences of ten Nordic actors with experience from renewable energy collaboration in Russia.

Russia's comparative advantages and disadvantages

Again utilizing the case study of the Nordic countries, the book maps some of the areas within renewable energy where Russia has particular strengths or weaknesses. In Russia, commercial, sociological and political approaches to renewable energy have enjoyed very low priority – and that has an important implication for international cooperation with Russia. The complementarities are evident: strong Russian basic research in the natural sciences complements the often strong skills of foreign actors in social science, commercialization and marketing. In the mapping of Nordic–Russian complementarities, hydrogen and solar power – both of which are high-tech fields where it is an advantage to draw on strong basic science – emerge as the best match of Russian and Nordic strengths in renewable energy. This may be the case for many other countries as well.

While acknowledging the potential of Russia's natural resource base and large population, foreign actors should be aware of the current obstacles to producing and selling renewable energy in Russia. At present it may make more sense to engage in projects oriented towards the export of materials, equipment and/or energy *from* Russia, than the production of renewable energy for the Russian market. On the other hand, although establishing renewable energy production in Russia may not yet be profitable in the short term, it could be worthwhile for actors wanting to position themselves for the future. In the long term, Russia could prove an exciting market for renewable energy.

As to cooperation of a more aid-oriented nature, the most productive areas are those related to changing policy in order to facilitate growth in the use of renewable energy. One important thing foreign actors could do is influence Russian climate policy. Transferring international technology for windmills, small hydro and the like makes little sense as long as the regulative framework and attitudes are not in place for implementing such technology.

Our understanding of 'cooperation'

Cooperation can take many forms and have different objectives – among them, aid, economic self-interest, environmentalism or joint scientific benefit. In this book we touch on activities that have a variety of objectives, but ultimately our interest is in cooperation with Russia that can generate mutual benefits in science and innovation through complementarities. This can be either through the joint development of scientific knowledge, or through invoking complementarities between scientific knowledge, natural resources, capital, management skills and/ or markets. Mutual benefit is also emphasized in the EU's approach to cooperation with industrialized and emerging economies.[33] Collaboration on renewable energy can also have other consequences:

- Scientific cooperation in an area such as renewable energy is also a form of relatively uncomplicated 'low politics'. When other issues cloud relations with Russia, continued collaboration in an area like renewable energy can be a good way to maintain at least a minimum of relations. That is not a primary focus of this book, but stands as a possible benefit of the cooperation discussed here.
- Collaborating with Russia on renewable energy will also in many instances make a contribution to the global effort against climate change and help to slow the depletion of hydrocarbons.

We see these possible side-effects of cooperation on renewable energy as positive and significant, but subsidiary to the main objective of complementary mutual benefit in science and innovation. In many cases, the collaboration may have several different consequences at the same time. For instance, strengthening Russian climate policy will obviously benefit the global effort against climate change, but it will also strengthen renewable energy in Russia, which in turn will ease cooperation with Russia on renewable energy.

Empirical basis

This book is based on a broad range of data. Seven fieldwork trips have been carried out: five to Russia, and one each to Copenhagen and Stockholm to cover the Nordic case from the authors' base in Oslo.

Over 100 interviews have been carried out for this study. With some individuals we have had little more than a fleeting, informal conversation; others were subjected to fully structured interviews, while yet others were interviewed

---

33   EC. *The European Research Area: New Perspectives*, Green Paper. Brussels: ECf, 2007b, p. 21.

by telephone. Ten of the interviews were carried out as part of a sub-component of the project exploring the experiences of Nordic actors with personal experience of cooperation with Russia on renewable energy. In these ten interviews we used an interview guide, which can be found in the appendices. From the five fieldwork trips to Russia we have amassed a large collection of written material on renewable energy. Assuming that most of our readers do not read Russian, we have not emphasized these sources in the book. We have cited some of them, but used others more indirectly for our own orientation.

The book has been reviewed by three prominent external peer reviewers, and significant feedback has also been provided by the organization Nordic Energy Research. In addition, various parts of the book have been presented in six presentations at four different international conferences. For further details about the interviews, peer review and conferences, see the appendices.

Overview of contents

The next two chapters provide an overview of the conditions for renewable energy development in Russia. Chapter 2 examines Russia's power sector, first looking at the reform of the Russian electricity monopoly RAO UES, and then moving on to examine the market potential for renewable energy in Russia. The reform of RAO UES, completed on 1 July 2008, is the most important event in the development of the Russian power sector during the past 15 years, and is therefore essential to an understanding of the Russian energy sector. It has resulted in the break-up of the former RAO UES conglomerate and the formation of a series of smaller companies, RusGidro being the one that focuses on renewable energy. At the end of Chapter 2 we examine some potential niche markets for renewable energy in Russia, and find that remote northern settlements could be an important niche.

Chapter 3 gives an overview of Russia's educational, scientific and innovation systems. These are important in highlighting the foundations and background of renewable energy research and development in Russia. Here we present a ranking list of the top educational-scientific institutions on renewable energy in Russia, which should help non-Russian actors in their search for partners. A section on Russian research funding organizations is also included.

Chapter 4 maps Russia's solar power sector. Russia excels in both solar power and hydrogen technologies, both of which are of potential global importance. The solar power sector was selected for detailed study because it is the most scientifically and industrially mature of the two. We found that the Russian solar power cluster is spread across several locations in the country and includes many commercially successful companies. Some of these, such as Nitol, are large and poised for international expansion.

The next three chapters map some of the existing international cooperation with Russia. Chapter 5 focuses on EU–Russian cooperation and provides a short overview of international projects. Chapter 6 provides a similar overview of Nordic–Russian cooperation. We use the case of Nordic–Russian cooperation to highlight the complementarities that can be utilized and taken into consideration when developing collaboration with Russian actors. Our mapping of complementarities between the Nordic countries and Russia is intended to serve as a model for other countries seeking to identify meaningful collaborative ventures with Russian partners, as well as for Russian actors who are looking for foreign partners.

Chapter 7 offers a qualitative description of the experiences of ten Nordic individuals with personal experience of working on renewable energy in Russia. This chapter is intended to help international actors wishing to engage in Russia's renewables sector to see what cooperation already exists, and what experiences the pioneering Nordic actors have had. This information can be useful in trying to fill in the gaps, avoid overlaps and find a place for new actors.

In the concluding chapter, we summarize our findings and put forward a set of policy recommendations for policy-makers, researchers and research funding agencies seeking to develop renewable energy partnership with Russia. The book ends by drawing up three scenario sketches for the development of the Russian renewable sector, aimed at helping both Russian and international actors to see the possible trajectories for Russia's renewable energy sector and position themselves in relation to these trajectories.

Chapter 2
Russian Energy Markets: Liberalization and Niches for Renewable Energy

Understanding the current state of the Russian electricity sector is crucial for analysing the potential for renewable energy investments in Russia. This chapter focuses on the liberalization of the Russian electricity sector and its implications for renewable energy and energy efficiency. Russia's electricity sector has recently undergone a lengthy reform process that culminated in the dissolution of the vertically integrated monopoly RAO UES (Unified Energy Systems) on 1 July 2008. One objective of this privatization was to attract the investment necessary for implementing energy-efficiency measures and developing renewable energy in Russia.

This chapter examines how the reform affects the conditions for developing renewable energy sources and the prospects of energy efficiency in Russia. We analyse the design of the energy sector reform and examine how its outcome may influence the prospects for a more environmentally friendly electricity sector by increasing the share of renewable energy and improving energy efficiency. All in all, the reform solves some problems, but creates new ones. We argue that the continued subsidizing of natural gas will remain the key impediment to the development of renewable energy sources. The benefits granted to incumbent energy producers constitute another roadblock to renewables. Due to these problems it still seems premature to expect any large-scale replacement of energy production capacity in Russia in the course of the next five years. At the end of this chapter we have therefore included an analysis of a possible niche market where renewable energy might thrive without competing with other subsidized energy sources in Russia.

The chapter starts with some basic information about the markets for renewable energy in Russia on a more general level, before we move to a more detailed analysis of the electricity sector reform. The implications of the reform for energy efficiency and introduction of renewable energy sources are then presented, before we conclude the chapter by presenting a possible market niche for renewable energy: the Northern Freight system.

Markets for renewable energy

Research on renewable energy in Russia has tended to deal mainly with the natural resource base for renewable energy, or with technologies for specific forms of renewable energy. The market potential for renewable energy has received less attention. Several factors make Russia a potentially strong market for renewable energy:

- Russia has the world's best base of natural resources for renewable energy, which means that prices for these resources should be lower than in countries where they are scarcer (e.g. Germany or China).
- With 142 million inhabitants, Russia is the world's eighth largest country in terms of population. That in itself bodes well for its potential as a market for renewable energy.
- The economy and the purchasing power of the population have been growing fast since 1998.
- Because energy for transport was priced far too low in the Soviet command economy, the country's demographic distribution is not optimal, with many towns and settlements located at great distances from each other. On the other hand, this can offer opportunities for renewable energy produced locally and in isolated locations.
- As a northern country with a cold climate and a highly energy-intensive industry, Russia needs massive amounts of energy.
- 22 million Russian families have *dachas* (country cottages or cabins), many of them not connected to the central grids.[1]
- Russia has much to gain from reducing its subsidies for natural gas and electricity. Above all, decreased domestic consumption as a result of higher prices could free up resources for exports. The government and major companies therefore have a strong incentive to reduce subsidies and create more equitable conditions for renewable energy to compete with traditional sources of energy.
- Russia is by far the world's largest country, covering a total of 17 million km² and 11 time zones, making it difficult to unite the entire territory under single electricity and natural gas grids. An estimated 10 million people are not connected to the central grids.[2]
- It has been estimated that five million farms are not connected to the central grids.[3]

As noted in the introduction, there are clearly also some significant limitations to the emergence of markets for renewable energy in Russia. Firstly, subsidies for natural gas and electricity put renewable energy at a big disadvantage. Removing these subsidies is difficult: the population has become accustomed to getting such commodities at very low prices since the Soviet period; moreover, the people have

---

1   Bezrukikh, Pavel. 'Netraditsionnye vozobnovlyaemye istochniki energii'. *Teplovoy Energeticheskiy Kompleks*, no. 4 (2001): 31–45.
2   Bezrukikh 2001.
3   Bezrukikh 2001.

experienced great hardships in the post-Soviet period and are wary of losing further social goods. Although Russian democracy has many weaknesses and problems, it is clear that the government and elite do not want to risk becoming unpopular with large segments of the population.

Secondly, there are problems with the reliability of the data that underpin the market picture. We tried hard to verify or find updated data on some of the points above (number of *dachas* and farms not connected to central grids, etc.), but in vain. Statistics Russia (formerly Goskomstat) has become more closed than it was a few years ago, perhaps due to a combination of greater commercialization and increasing suspicions about security threats. RAO UES had data on those who were connected to its grids, but lacked data on those not connected. Some interviewees opined that most *dachas* are connected to the central grids, others disagreed with this view, but no reliable data exist. This lack of data is an obstacle to the development of renewable energy markets, since investors and innovators are forced to operate in uncharted territory.

Electricity sector reform

In recent decades, the electricity sectors of many countries have been reformed and liberalized. Chile experimented with market reform in 1987, while England and Wales followed in 1989. The Nordic countries established Nord Pool as the world's first multi-national exchange for trading electric power in 1996. These reforms follow a basic formula that includes restructuring, liberalization, privatization and deregulation. Although there are variations, the desire to change the organization of the electricity sector is commonly driven by a wish to make the sector more efficient by introducing multiple players and competition. Improvements are expected to include better rationalization of labour and fuel costs in power generation and procurement, superior investment decisions and allocation of risks, and enhanced customer service. Fereidoon Sioshansi also notes that in developing countries, reforms are commonly intended to attract investment to the power sector.[4] Russia is no developing country: but it has a power sector in desperate need of investments, and attracting investments is the primary goal of the reform.

A short note on the various terms for different forms of electricity sector reform is in place, to avoid confusion. Sioshansi and Pfaffenberger distinguish between restructuring, liberalization, privatization, corporatization and deregulation.[5]

---

4   Sioshansi, Fereidoon P. 'Electricity Market Reform: What Have We Learned? What Have We Gained?'. *The Electricity Journal*, vol. 19, no. 9 (2006): 70–83, p. 70.

5   Sioshansi, Fereidoon P. and Wolfgang Pfaffenberger. *Electricity Market Reform: An International Perspective*. Amsterdam: Elsevier, 2006, p. 41.

Restructuring, they write, is a broad term, referring to 'attempts to reorganize the roles of the market players, the regulator and/or redefine the rules of the game, but not necessarily "deregulate" the market'. *Liberalization* refers to 'attempts to introduce competition in some or all segments of the market, and remove barriers to trade and exchange'. *Privatization*, perhaps needless to say, refers to 'selling government-owned assets to the private sector'. This, they point out, has been done in most countries that have embarked on market reform. *Corporatization* is the attempt 'to make SOEs [State Owned Enterprises] look, act and behave as if they were for-profit, private entities'. This includes the introduction of competition between companies that belong to the same shareholder. Lastly, *deregulation* refers to 'removing or reducing sector-specific regulation and subjecting the ESI [Electricity Supply Industry] to monitoring by the anti-cartel authority'. However, these authors also note that no electricity market has been, or can be, fully deregulated, and that there is agreement on the need for specific regulation of transmission and distribution of electricity.

According to the categorization supplied by Sioshansi and Pfaffenberger, Russia's current electricity sector reform is most correctly referred to as both restructuring – since it involves a redefining of the rules of the game – and liberalization, as competition is introduced in some segments of the market, as well as privatization, since the assets of the state-owned monopoly are sold to the private sector. However, Viktor Balyberdin's reference to the Russian reform as a 'complete deregulation'[6] seems somewhat extreme in terms of Sioashansi and Pfaffenberger's categorization, since the monopoly is retained in transmission and distribution. We return to this point later.

The degree of success of reforms in electricity markets varies greatly. In nearly all cases, the initial reforms have led to 'reforms of the reforms', and 'hybrid markets' have been the result of several reforms.[7] It is natural to learn from previous efforts when embarking on a new reform. Although the Russian reformers have not stated clearly which reform example they aim to follow, the case of California has been mentioned as an example of what to avoid.[8] Also in the literature on electricity market reforms, the reform in California is frequently cited as a deterring example, due to the 2000/01 electricity crisis that followed the restructuring of the electricity market there.[9] The California crisis is referred to as a 'market

---

6   Balyberdin, Victor. 'Russian Power Market : Structure, Development, Prospects'. Presentation held at the Energy Forum conference *Investing and Financing Renewable Energy in Russia*, Stockholm, Sweden, 15–16 April 2008.

7   Sioshansi 2006, pp. 70–83.

8   Shuster, Simon. 'Gazprom, Interros Ready to Carve Up Power Industry'. *St Petersburg Times*, 13 February 2007. http://www.sptimes.ru/index.php?action_id=2&story_id=20354 [accessed 21 June 2009].

9   See e.g. Sioshansi 2006.

meltdown'. New generation capacity is badly needed in Russia, but there exists considerable uncertainty about who will build this and how investments will be recovered. Pending clarification, today's system resembles the old command-and-control system from before the restructuring.[10] On the other end of the scale is the Nordic model for electricity market reform, which stands out as the best example of a successful reform.[11] Four success factors seem to have been decisive: The simple but sound market design made possible by a large share of hydropower; the dilution of market power created by the integration of four national markets into a single Nordic one; firm political support for a market-based electricity supply also in situations of market distress; and the firm voluntary informal commitment by the Nordic power industry to public service.[12] On the other hand, there are doubts as to whether the success in the Nordic countries can readily be transferred to other countries. Only two of the four success factors are seen as transferable (the second and third), while the rest are largely country-specific.[13]

These examples from countries that have had a more developed market economy, a better regulatory framework in place and more transparent business environments than Russia when embarking upon electricity sector reforms, clearly show the pitfalls in this field. This is not to say that Russia cannot succeed with the reform, or that reform in itself is to be avoided: but it is vital for Russia to learn from mistakes made elsewhere.[14] Even though Russian energy reform proponents have explicitly sought to avoid the mistakes made in California, critics have remarked that the Russian reform model might actually end up repeating all those same errors. Russell Pittman claims the Russian electricity reform is likely to create significant levels of market power and even monopoly power under peak-load conditions in certain regions.[15] This is inexorably linked to geographic factors in the world's largest country. Due to the poor conditions and low capacities of inter-regional high-transmission linkages, there are likely to be six distinct regional wholesale electricity markets in Russia, not one national market. We will return to how realities on the ground in these regional markets may well contribute to creating monopoly-like or cartel-like situations.

10　Sioshansi 2006, p. 76.

11　Sioshansi 2006, p. 72.

12　Amundsen, Eirik, Lars Bergman and Nils-Henrik M. von der Fehr. 'The Nordic Electricity Market: Robust by Design?', in *Electricity Market Reform: An International Perspective*, edited by Fereidoon Sioshansi and Wolfgang Pfaffenberger. Amsterdam: Elsevier, 2006, pp. 145–70, p. 169.

13　Amundsen, Bergman and von der Fehr 2006.

14　See Xu, Yi-Chong. *Electricity Reform in China, India and Russia: The World Bank Template and the Politics of Power*. Northampton, MA: Edward Elgar, 2004 for a discussion of the problems inherent in pushing a policy reform agenda for the electricity sector that requires more sophistication in governance and regulatory mechanisms than is available.

15　Pittman, Russell. 'Restructuring the Russian Electricity Sector: Re-creating California?'. *Energy Policy*, vol. 35, no. 3 (2007): 1872–83, p. 1873.

If the reform does not succeed in creating competition, it will have failed to achieve one of its main goals. That would be a major problem for the development of renewable energy sources in Russia. Before describing the design of the reform and its implications for the possibilities of developing renewable energy sources, we need to take a look at Russia's energy strategy and policy, and the role played by renewable energy sources in the strategy.

Russian energy policy: Room for renewables?

Russian great-power ambitions, which became increasingly evident under the Putin presidency, have been linked to the country's role as an energy superpower. These ambitions rely to a large extent on replacing Soviet Russia's former military might with economic power tools.[16] Russia has set for itself the ambitious goal of doubling its GDP within a decade – and that will require stable, reliable and inexpensive electricity supplies for the country's large companies.

Estimates of Russia's energy efficiency potential demonstrate that saving energy could in fact prove to be the country's greatest source of increased energy for domestic consumption and exports.[17] As described in the previous chapter, the economy is in itself highly energy-intensive, and the use of energy can be characterized as generally wasteful. That today's Russian economy is wasteful in energy is also acknowledged in the Russian energy strategy,[18] which sets relatively ambitious goals for reducing the energy intensity of the country's economy.[19] Figure 2.1 shows Russia's energy-efficiency potential in various sectors.

Although one may question how firm the link is between this ambitious goal and the actual day-to-day policy-making relevant for achieving it, there is at least a clear intention at the rhetorical level in the Russian Energy Strategy of doing something about the country's strikingly inefficient use of energy. However, the same does not go for the underdeveloped renewable energy sector: these energy

---

16   Lo, Bobo. 'Evolution or Regression? Russian Foreign Policy in Putin's Second Term', in *Towards a Post-Putin Russia*, edited by Helge Blakkisrud. Oslo: NUPI, 2006, p. 63.

17   Merle-Béral, Elena. 'The Wider Perspective: Russia's Energy Scene'. Presentation at the conference *Renewable Energy in Russia: How Can Nordic and Russian Actors Work Together?*, Oslo, Norway, 8 May 2008.

18   Ministry of Energy of the Russian Federation. *Summary of the Energy Strategy of Russia for the Period of up to 2020*. Moscow: Ministry of Energy of the Russian Federation, 2003.

19   The *Energy Strategy* proposes using a wide range of methods, including 'a structural rebuilding of the economy in favour of low power consuming manufacturing industries, knowledge industry and human services [...]'. Ministry of Energy of the Russian Federation 2003, p. 7.

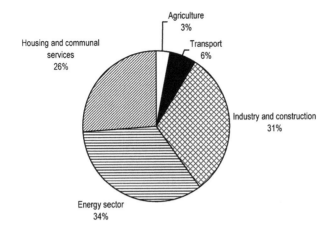

Figure 2.1   Russia's energy-saving potential by sector[20]

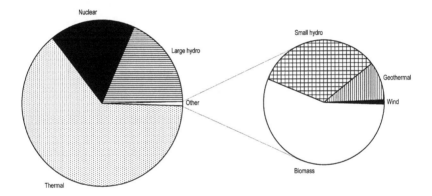

Figure 2.2   Share of renewables in Russian electricity generation[21]

sources receive only passing mention in the Energy Strategy. Figure 2.2 shows the share of renewable energy sources in electricity generation in Russia.

The Russian electricity sector before the reform

Russia's state-controlled electricity monopoly had its origins in Lenin's ambitious effort to bring electricity to the people of the newly-founded Soviet

20   Source of data used to compile figure: Merle-Béral 2008.
21   Source of data used to compile figure: Merle-Béral 2008.

Union.[22] After the break-up of the Soviet Union, the initial period of privatization of the Russian economy brought partial privatization of the Russian part of the Soviet electricity sector when the United Energy System from the Soviet period was transformed into a single joint-stock company with the federal government as its principal shareholder. The vertically integrated state monopoly RAO UES was established as a transitional structure in 1992, pending more comprehensive reform of the electricity sector.[23] However, further privatization of the energy sector in the 1990s was marred by the conflict between the Yeltsin administration and the parliament, and was abandoned upon the financial crisis in 1998.[24]

In the early 2000s, reform plans began to take shape. At the onset of the reform, RAO UES was generating approximately 70 per cent of Russia's electricity production. The base load of 30 per cent is accounted for by nuclear power plants, which will remain under the control of the Ministry of Atomic Energy (MINATOM) also after the reform. In addition to generating capacity, RAO UES has also owned the central dispatch administration, the Federal Network Company, 36 power plants, around 57 R&D institutes, and has had stakes in more than 70 construction, maintenance and service companies, as well as 2.5 million km of power lines.[25] When the reform began, the government held slightly over 52 per cent of the RAO UES stock, and foreign shareholders owned approximately half of the privately-held shares. Figure 2.3 shows the structure of the Russian electricity sector before the reform.

RAO UES has to this day suffered continuously from what Leon Aron calls 'congenital Soviet maladies: distorted prices, waste and decay'.[26] Prices for electricity have been set on a yearly basis by the Federal Energy Commission and regional energy commissions. These prices have by no means covered the costs of production, and this has led to a situation of endemic waste. Between 30 and 40 per cent of Russian energy is lost in production, transportation, transmission or inefficient consumption.[27] Cross-subsidies have meant that residential customers have paid a significantly lower price than either agriculture or industry. In addition, the company's assets are 30 years old or more, and approaching the end of their planned operational life, so there is a desperate need for investments – in the range

---

22   Wengle, Susanne. 'Power Politics: Electricity Sector Reforms in Post-Soviet Russia'. *Russian Analytical Digest*, no. 27 (2007): 6–9.

23   Tompson, William. *Restructuring Russia's Electricity Sector: Towards Effective Competition or Faux Liberalisation?*, OECD Economics Department Working Papers No. 403, 2004.

24   Aron, Leon. 'Privatizing Russia's Electricity'. *Russian Expert Review*, no. 4 (2003): 9–17, p. 12.

25   Tompson 2004, p. 5.

26   Aron 2003, p. 10.

27   Aron 2003, p. 11.

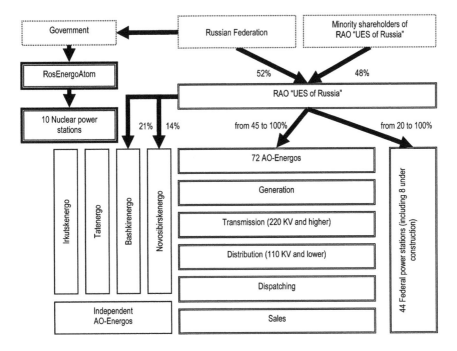

Figure 2.3   The Russian electricity sector before reform[28]

of hundreds of billions of dollars. Obviously, attracting investment on that scale is impossible in a loss-making and non-transparent business environment.[29]

Demise of a monopoly

Due to the specificities of the Russian electricity sector, and as a result of efforts since the demise of the Soviet Union to transform the Russian economy, the reform was conceived as a plan to make the electricity sector competitive, transparent and profitable. Anatoliy Chubais has been the CEO of RAO UES since 1998. He was a key figure in the group of 'young reformers' during the privatization of state property in Russia in the early 1990s, and earned his nickname as the 'privatization tsar' for his role in the privatization programme. Chubais, who has a strong history of commitment to liberal reform, has introduced significant changes in the running of the company. In 1998, only one-third of its settlements with generators and consumers were made in cash. Non-payment prevailed – in-kind

---

28   Source of data used to compile figure: RAO UES (United Energy Systems of Russia) webpage. http://www.rao-ees.ru [accessed May 15, 2008].

29   Aron 2003.

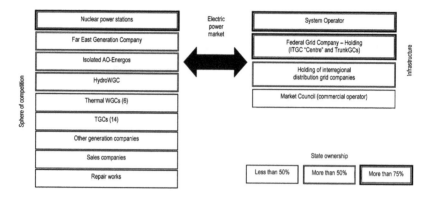

Figure 2.4   Structure of the Russian electricity sector after reform[30]

payments or settlements within a complex and corrupted system where the state
subtracted electricity bills from its debts to industrial suppliers. After RAO UES
was allowed in 2000 to disconnect non-paying customers and Chubais succeeded
in getting a ban on all non-cash settlements from January 2001, the proportion of
cash payments rose to 92 per cent in 2001 – an impressive accomplishment that
shows a firm commitment to change.[31]

As mentioned, the California example has been a deterrent. Perhaps more
interestingly, there seems to have been no clear role model for the Russian reform.
Yet there can be no doubt as to who the main propagator of reform has been: CEO
of RAO UES, Anatoliy Chubais. His orchestration of his own company's break-up
was supported by President Putin, who made a U-turn in his position on the reform
after entering office, and signed the plan into law on 31 March 2003. Opponents
of the plan ranged from government bureaucracies unwilling to lose their control
over electricity prices and distribution, through both the liberal and communist
opposition in the Duma, to governors fearing their loss of control over local
energos and the various industries dependent on cheap subsidized electricity.[32]

Figure 2.4 shows the design of the reformed electricity sector. On the left are
infrastructure assets that will continue to have a high level of state ownership and
control. The newly established not-for-profit organization Market Council (*Soviet
rynka*) is composed of organizations of the commercial infrastructure of the
wholesale market, including a trade operator. The Market Council is vested with
the dominant power to regulate the industry. On the right are the branches where
more competition is introduced. Nuclear power stations remain under the control

---

30   Source of data used to compile figure: RAO UES web page.
31   Aron 2003.
32   Aron 2003.

of the Ministry of Atomic Energy, MINATOM, while generation companies, sales companies and repair works are subject to competition and less state control.

The state will be the majority owner of the Federal Grid Company and the System Operator. This isolates one part of the electricity sector (high-voltage transmission lines and low-voltage distribution lines), which is kept as a natural monopoly, while competition is introduced in generation and retail. The government will also regulate prices. The separation between natural monopolies and sectors where competition is introduced is common in electricity sector reforms. In the literature on electricity sector reforms there is agreement on the need for 'monopolistic bottlenecks of the market in transmission and distribution'.[33] The continued monopoly has nevertheless been interpreted as something specifically Russian, and as reflecting the view of reform opponents, who seek to classify electricity as a 'strategic sector' and not merely an infrastructure sector.[34] However, this view fails to take into account the considerable literature on energy sector reforms, which emphasizes precisely the need to separate transmission and distribution, as less competitive parts of the electricity market, from the competitive market in retail and production.[35] In fact, the Russian reform design resembles what could be called the 'textbook model' for restructuring and competition.[36]

Nevertheless, there is concern that the reform will not succeed in creating competition. For the development of renewable energy sources this is particularly worrisome, as lack of competition and regional monopoly-like situations will be a major impediment to the introduction and development of renewables. There is also concern that the reform fails to establish the necessary regulatory mechanisms with the competence needed to open up for renewables. Also, there are non-reform-related hurdles to the development of renewable energy sources which the reform does not change, and which could prove the greatest impediment to creating competition between non-renewable and renewable energy sources: specifically, the continued existence of gas subsidies.

---

33   This is how Sioshansi and Pfaffenberger refer to the notion that electricity markets can not be fully deregulated (Sioshansi, Fereidoon P. and Wolfgang Pfaffenberger. *Electricity Market Reform: An International Perspective*. Amsterdam: Elsevier, 2006, p. 41).

34   Wengle 2007, p. 8. Aron (2003) also describes the reform design as a 'contradictory compromise' between reform proponents and opponents.

35   Transmission and distribution are parts of the market where there seems to be agreement on the need for a greater degree of control and monopoly despite the introduction of competition within retail and production. See Sioshansi and Pfaffenberger 2006.

36   Littlechild, Stephen. 'Foreword: The Market Versus Regulation', in *Electricity Market Reform: An International Perspective*, edited by Fereidoon Sioshansi and Wolfgang Pfaffenberger. Amsterdam: Elsevier, 2006, p. xviii.

Regional electricity markets

As mentioned, Russell Pittman has warned that the Russian electricity sector reform will repeat the mistakes from California. More specifically, he is concerned that the Russian electricity sector will create grounds for anti-competitive behaviour.[37] Pittman voices serious doubts as to whether the reform will be able to create competition in generation markets. Through a detailed survey of the structure of the regional markets, he finds that the reform is 'likely to create significant levels of market power and even monopoly power under peak-load conditions in particular regions – market and monopoly power that could be exercised with great harm to Russian economic welfare'.[38] This is because the poor conditions and low capacities of inter-regional high-transmission linkages mean that there will be six regional wholesale electricity markets, and not one national market.[39] The thermal generation facilities now owned by RAO UES will be sold off to six new private companies of roughly equal size. In addition there will be 14 smaller, territorial generation companies owning and operating the smaller generation plants. The six large generation companies will encounter each other in 'multiple geographic markets in repeated interactions over time'.[40] The primary concern with regard to creating competition is whether these companies will be able to learn from their experiences and coordinate their interactions so as to maximize prices and profits, rather than competition and efficiency in particular regional markets. This will not be limited by the rule of allowing one company to own maximum 35 per cent of the generation capacity in any wholesale price zone.[41] The markets all appear at least moderately concentrated. Pittman finds specific reason for concern in this regard in the Volga region (where he describes a two-firm oligopoly at best, or a Tatenergo virtual monopoly at worst), the Central region, and the Northwest region (both these markets are so concentrated that there is a great possibility of coordinated behavior).[42]

Implementation of the energy sector reform

It is too early to assess the results of the Russian electricity sector reform. Much is likely to depend on not just the design of the reform, but its implementation as

---

37   Pittman, Russell. 'Restructuring the Russian Electricity Sector: Re-creating California?'. *Energy Policy*, vol. 35, no. 3 (2007): 1872–83.

38   Pittman 2007, p. 1873.

39   The regions are identified by RAO UES as the Northwest (including St Petersburg), Central (including Moscow), South, Volga, Urals and Siberia. A potential seventh regional market, Pittman notes: the Far East, is so fragmented that it is likely to be composed of multiple smaller regional geographic markets (2007, p. 1873).

40   Pittman 2007, p. 1874.

41   Pittman 2007, pp. 1873–4.

42   Pittman 2007, pp. 1877–80.

well. Chubais seems satisfied with the reform design, which, he has opined, will create 'the world's most advanced electricity market'.[43] Experts also agree that the new design of the Russian electricity market *is* good – but certain problems and challenges are expected when it comes to implementation.[44] The balance the reform aims to strike between control and liberalization might be vulnerable to political manipulation and corruption, which could distort the market and discourage investors. For example, in transmission some suppliers could be favoured over others through beneficial contacts with the right officials, in turn disadvantaging other suppliers. This would disturb the competition and impede the stimulation of a market situation – which is the aim of the reform. For renewable energy resources, as a newcomer to the market, this would probably be a major impediment.

At present there seems to be considerable interest in the Russian energy sector on the part of both Russian and foreign investors. From abroad, companies like EON, RWE and Japan's Mitsui have shown an interest in acquiring assets. This is a large market in which many apparently want to have a share. Importantly, the balance of payments seems likely to be favourable in the next 10 years, giving good reason to expect that investments will be profitable. The lion's share of domestic investments comes from the state-controlled gas monopoly Gazprom. The company has secured some assets that have been up for grabs in connection with the break-up of RAO UES, and seems interested in electricity sector assets in order to be able to step up the use of coal for domestic power production, so as to increase exports of gas.[45]

The role of Gazprom in the Russian energy market deserves separate treatment. The gas monopoly is commonly viewed as the prolonged arm of the Kremlin. Today, with Dmitry Medvedev as head of the company's management board and simultaneously President of Russia, the blurring of what are government policies and what are company policies is likely to continue. It remains to be seen what greater Gazprom control of electricity sector assets will mean, but this does not seem to bode well for the development of renewable energy sources. Gas is of course the main focus of Gazprom, and gas stands for 53.6 per cent of total primary energy supply in Russia.[46] Oil and gas exports are responsible for approximately

---

43   *Russian Courier*. 'Russia Launches Advanced Energy Market', 1 September 2006. http://www.russiancourier.com/en/news/2006/09/01/59466/ [accessed 5 May 2008].

44   Interview with senior banker in the Department of Power and Energy in the European Bank for Reconstruction and Development, Moscow office, 29 January 2008.

45   Kramer, Andrew E. 'Gazprom Moves into Coal as Way to Increase Gas Exports'. *New York Times*, 27 February 2008. http://www.nytimes.com/2008/02/27/business/worldbusiness/27coal.html?ex=1361854800&en=17898f35b32aa09b&ei=5088&partner=rssnyt&emc=rss [accessed 29 February 2008].

46   IEA. *Renewables in Russia: From Opportunity to Reality*. Paris: IEA, 2003. http://www.iea.org/textbase/nppdf/free/2000/renewrus_2003.pdf [accessed 29 February 2008].

60 per cent of Russia's export revenues,[47] and increasing gas exports can be expected to be a main goal for Gazprom. Gas for domestic consumption has been heavily subsidized – a major issue in Russia's negotiations for WTO membership. In these negotiations, Russia has committed itself to cut its gas subsidies by 2011. However, according to a May 2008 announcement from the Ministry of Economic Development and Trade, domestic gas prices will not reach the European market level before 2014/15, apparently pushing the abolition of subsidies even further into the future.[48] This in itself distorts the market and is an impediment to the profitability of other energy sources, justifying the charge that there is no real competition between non-renewable and renewable energy sources in Russia today.

Gazprom is however not completely blind to climate issues, and has been adjusting somewhat to the new realities created by the Kyoto Protocol. For example, the company is engaging in carbon trade, and would seem to acknowledge the business opportunities inherent here.[49] Carbon capture and storage could therefore be a potential area for international cooperation with Russia in the future.

Implications of reform for renewable energy sources

While energy efficiency is acknowledged as a goal in itself in the Russian Energy Strategy, the potential of renewable energy sources is, despite the proven potential, not equally recognized as important for reaching the country's goals. This may partly be due to the fact that the Energy Strategy was published back in 2003, when renewables were accorded far less attention internationally as well. Renewable energy sources are not listed separately, but receive only a passing mention with the unspecific goal of increasing their share.[50] The reasons behind the lack of development of this tremendous potential are manifold, ranging from political and bureaucratic, to financial and technical factors. The reform of the energy sector can be expected to alleviate some of these problems, but not all. Hydro OGK was mandated to draft a law on renewable energy, planned as a set of amendments to the law on restructuring of RAO UES, with tax breaks and green subsidies to renewable energy. The draft, however, was not approved. It still seems likely that in the short to medium term these energy resources will be overshadowed by Russia's petroleum resources, which are central to the country's Energy Strategy

---

47   Energy Information Administration. 'Russia Energy Profile'. http://tonto.eia.doe. gov/country/country_energy_data.cfm?fips=RS [accessed 9 June 2008].

48   Hemscott.com. 'Russian Domestic Gas Prices to Remain below European Export Level to 2014–15', 27 May 2008. http://www.hemscott.com/news/static/tfn/item. do?newsId=64497524037005 [accessed 7 June 2008].

49   Medvedev, Alexander I. 'Growth Solutions for the Low Carbon-Economy'. Keynote speech, *Global Leadership and Technology Exchange* meeting, Moscow, 4 July 2008.

50   Ministry of Energy of the Russian Federation 2003.

of 2003. Lacking real chances to compete on equal terms may also hinder foreign investors in fulfilling the constructive role they would like to play in renewable energy development, under different conditions.

The monopoly that will remain in transmission and distribution poses major challenges to regulation. The success of the reform will hinge on the ability to create competition among suppliers, all dependent on equal access to the grid. Wherever there is a limited good, there is room for conflict and corruption; the result might be a situation where corruption grows out of suppliers' relations to the grid company. Producers of renewable energy sources, as newcomers in the market, are less likely to have the contacts and necessary know-how for manoeuvring around the system. How might this affect the competition among producers? It is obvious that the value of a generating asset will be hugely affected by access to the grid and by investments that are made (or not made) in the grid in question.[51] This is a call for establishing truly independent regulatory mechanisms, and points to the necessity of building incentives and the right institutions to enable Russia to realize its potential in renewable energy sources and energy efficiency.[52]

Another impediment to the introduction of new energy sources embedded in the reform is that incumbent generators receive payments based on their capacity to produce energy, regardless of whether or not they actually *sell* energy. This could work as a disincentive to invest and impede new entries to the market,[53] thus hindering the establishment of renewables production. History has shown that the absence of institutional frameworks for independent power producers has been a barrier to renewable energy in Russia.[54]

However, the most crucial issue for the development of renewable energy sources in Russia is not addressed at all by the reform of the power sector: that is the absence of a level playing field for non-renewable and renewable sources of energy. Although the subsidies for domestic gas use are set to diminish by 2011 as part of Russia's adaptation to WTO rules, this is perhaps the key impediment to the development of renewable energy sources in the short term. At present, gas subsidies are not balanced by possible other financial incentives to producers of renewable energy sources. A package of various stimuli is therefore necessary to promote the development of renewables in Russia. This might take the form of 'green subsidies'

---

51   Tompson 2004, p. 19.

52   Golub, Alexander and Elena Strukova. 'Russia and the GHG Market'. *Climatic Change*, no. 63 (2004): 223–43, pp. 240–41.

53   Tompson 2004, p. 17.

54   Martinot, Eric. 'Energy Efficiency and Renewable Energy in Russia: Transaction Barriers, Market Intermediation, and Capacity Building'. *Energy Policy*, vol. 26, no. 11 (1998): 905–15, p. 910.

and tax breaks for renewable energy sources, green certificates, price incentives or benefits granted to renewables in a domestic emissions trade market.[55]

Energy efficiency: A likely outcome of the reform?

Saving energy is recognized as a necessity in Russia's Energy Strategy of 2003. Increasing the efficiency of power utilities is one of the two primary goals of the power sector reform as stated in the Concept of RAO UES strategy for 2003–2008. There seems to be at least some political will to implement energy-efficiency initiatives, so a strong commitment and improvements in energy efficiency should be expected. To the degree that the reform succeeds in attracting new investors to the sector – be it as new owners of old power plants or in retail – the investments are likely to emphasize energy efficiency in all stages. The benefits of getting the most out of the energy produced are clear to all. In moving towards better energy efficiency, actors can also choose to implement only cost-efficient measures, of which there are many, and this means that investments are guaranteed to be profitable.

As part of the reform process, electricity prices are also set to increase and come closer to reflecting the actual production costs. This is a strong incentive for all consumers, both industrial and residential, to reduce energy consumption. However, to reap these great possibilities for energy efficiency, there is also a need to implement metering and other technical measures.

Results of the reform

The reform of the Russian electricity sector seems capable of attracting investments – which is, after all, its main goal. Some of this may go into the development of renewable energy sources, although major challenges remain to be overcome before this can become a booming sector of the economy. We have also noted the well-founded concern relating to what kinds of investors are attracted to the Russian electricity sector. Gazprom's role as a major investor would not seem to bode well with regard to the development of renewable energy sources.

Gas subsidies have been identified as the key impediment to the development of renewables in Russia. In the short term, such subsidies will continue to impede the development of renewables unless other incentives or stimuli can be implemented as a counterweight – e.g., green certificates, tax breaks or benefits from emissions trade. In addition, new structural challenges are created by the reform, like the

---

55   Interview with senior banker in the Department of Power and Energy in the European Bank for Reconstruction and Development, Moscow Branch, 29 January 2008.

benefits granted to existing energy producers, and the absence of independent regulation of grid access. For the development of renewable energy sources, it therefore seems that while the reform of the electricity sector may solve some problems, it will probably create others.

The prospects are brighter for energy efficiency measures, if only because wasting produced energy is extremely unattractive, and the benefits of getting the most out of what is produced are obvious. Also there are plenty of cost-efficient energy efficiency opportunities ready and waiting. The investments attracted to the Russian energy sector through the reform are therefore likely to bring about significant efficiency gains in production, while there are still structural factors that need to be addressed to enable necessary improvements in transport and consumption. With the continued strength of the natural monopoly in gas and the subsidized gas prices, it seems unlikely that Russian perceptions of energy resources will shift significantly in the direction of recognizing them as finite. On the other hand, to the degree that energy prices rise, this is likely to act as a strong incentive to limit energy consumption, for residential and industrial consumers alike.

To sum up, the Russian electricity sector reform can stand as an example of the limits to what the market can do with regard to changes towards more environmentally-friendly energy systems. The Russian government needs to develop a clearer strategy and establish institutions that can ensure the development of renewable energy sources and the implementation of energy efficiency measures on all levels.

Our analysis of the Russian electricity sector reform has identified some of the challenges to renewable energy development existing in the Russian energy sector. We will now move to focus on a possible market niche where renewable energy might well thrive without competing with other subsidized energy sources.

Possible low-hanging fruit: Northern Freight

Northern areas with harsh climatic conditions make up around 60 per cent of Russia's territory. Population density is low, less than one person per km^2, making it prohibitively expensive to connect most of these areas to the central grids. Over 6600 diesel-powered plants with a total capacity of 3.3 GW are in use in these parts of the country, and they require around 2 million tonnes of diesel annually.[56]

Severny zavoz or 'Northern Freight' is a system of subsidized goods and transport for Russia's Arctic, sub-Arctic and other remote locations with harsh

---

56   Marchenko, O.V. and S.V. Solomin. 'Efficiency of Wind Energy Utilization for Electricity and Heat Supply in Northern Regions of Russia'. *Renewable Energy*, vol. 29, no. 11, (2004): 1793–809, p. 1796.

climatic conditions.[57] Most of the goods transported are foodstuffs and fuels for heating, electricity generation and transportation. The system was established during the first decades of the Soviet period in order to help populate Russia's vast Arctic and sub-Arctic expanses, a major objective for Stalin and Khrushchev. Northern Freight was one of the three pillars of the Soviet demographic push into the Arctic, the other two pillars being forced migration to the *gulags* and special socio-economic benefits like higher wages, longer holidays and earlier retirement for those who went to live and work in the northern parts of the country.

Since the collapse of the Soviet Union, the Northern Freight system has been maintained as a matter of tradition and in order to avoid the socio-economic and political destabilization of settlements in the North. In connection with the financial and monetary crash in 1998 there was some discussion of scrapping the Northern Freight system and abandoning many of the remote settlements, but in practice the system has continued.[58]

The centrepiece of the Northern Freight system is its long-distance transport of fossil fuels, mainly diesel and coal, to remote settlements. Most spectacular is the transport of diesel, variously carried out by river tanker, biplane, helicopter or tracked vehicle to isolated settlements. Needless to say, this is a costly endeavour. Rental of the Mil Mi-8 or larger helicopters needed for bulk freight costs at least 1,500 USD per hour. In some cases, ships carrying coal or diesel are accompanied by nuclear icebreakers, at a cost of 600,000 roubles per day, not counting the subsidies for nuclear fuel.[59] In some parts of the Russian Federation, such as Altai and Tuva on the Central Asian border and Kamchatka in the Far East, over half of the regional budget is spent on fuel that is included in the Northern Freight system.[60] It has been estimated that in the Soviet period, 3.5 per cent of GDP was spent on various Northern benefits, including Northern Freight.[61] Information is not available concerning the total cost of Russia's Northern subsidies today, but in current GDP terms 3.5 per cent would correspond to USD 34.6 billion.[62] In 2007, the entire Northern Freight included 17.3 million tonnes of different products, an increase of 2.3 per cent from 2006. Out of this, petroleum products made up

---

57   *Severny zavoz* is also variously translated as 'Northern shipment' or 'Northern deliveries'. See e.g. Heleniak, Timothy. 'Migration and Restructuring in Post-Soviet Russia'. *Demokratizatsiya*, (Fall 2001): 1–18.

58   Heleniak 2001.

59   Nikitin, Oleg. 'Kak perevozitsya neft'. *Tekhsovet*, vol. 33, no. 2, 5 February 2006: 2–4, p. 4.

60   Bezrukikh 2001.

61   Heleniak 2001, p. 10.

62   World Bank in Russia. *Energy Efficiency in Russia: Untapped Reserves*. Moscow: World Bank, 2008, p. 341.

2.7 million tonnes, and coal 1.6 million tonnes.[63]As an example of what the Northern Freight of carbon-based fuels involves for one region, we can take the sparsely populated Yamal Peninsula in northwest Siberia with a population of 16,000.[64] In 2007, this population alone received:[65]

- 200,000 tonnes of petroleum products
- 20,500 tonnes of coal
- 40,000 m³ of firewood

According to Douraeva, combined wind/diesel installations can reduce the fuel needs in remote locations by up to 80 per cent.[66] According to Kreutzmann, both wind and solar installations could replace the diesel brought in through the Northern Freight system.[67] Although this would not fully obviate the need for fuel from the outside world, it could cut it significantly, and North Freight expeditions could be reduced to every second or third year.

Remote settlements in the North have the advantage of being able to use wind power for both electricity and central heating. Any surplus electricity can be transformed into hot water which can be stored in insulated tanks for a while until it is distributed through the central heating system.[68]

Perhaps the most significant advantage of these remote Northern settlements as a market for renewable energy is precisely the fact that they are not connected to the central electricity and natural gas grids. As a consequence of their isolation, renewable energy in these locations will not compete with subsidized natural gas and nuclear power, and the owners of renewable energy power plants need not worry about the technical and legal details of connecting to the grids.

---

63   Rosbalt News Agency. *Mintrans Raportuet o Vypolnenii Severnogo Zavoza*, press release, 28 September 2007. http://www.rosbaltnord.ru/2007/09/28/417963 [accessed 4 March 2008].

64   This population figure for Yamal is from 1993 and is therefore outdated. Since about half the population consisted of Slavic and other groups that moved north during the Soviet era and have undergone net emigration during the post-Soviet period, actual figures today are unlikely to be any higher. Krivitsky, Sergey and Alexander Tsvetinsky. 'Oil and Gas Exploration on the Arctic Shore'. *Proceedings of the Eleventh International Offshore and Polar Engineering Conference*, Stavanger: International Society of Offshore and Polar Engineers, Stavanger, Norway, 17–22 June 2001, pp. 661–4, p. 662.

65   Ioffe, Olga. 'Severny gorizont Rossii'. *Tekhsovet*, vol. 33, no. 2 (5 February 2006): 7–12, p. 8.

66   IEA 2003, p. 59.

67   Kreutzmann, Anne. 'The Smell at the End of the World: Nitol Wants to Produce Silicon in Siberia'. *Photon International*, no. 11, November (2007): 30–47.

68   Marchenko and Solomin 2004, p. 1795.

Obstacles

Despite these aspects that can make remote Northern settlements attractive as
a breakthrough market for renewable energy, there are also some significant
obstacles. One of the main drivers for the continuation of the Northern Freight
is corruption. Although it is widely understood that most of the numerous central
heating and electricity generation plants in the country are highly inefficient, it is
difficult for the central authorities to get an overview over exactly how inefficient
specific utilities are – especially those situated in remote locations.

This leaves considerable scope for misappropriation of fuel brought in by Northern
Freight. Since no one knows exactly how much is needed, it is also difficult to know
how much is illegally diverted. However, it is clear that many of the resources provided
through the Northern Freight system find their way into private hands, and then to
local or regional markets. In some locations with largely non-monetized economies
this phenomenon is so widespread that diesel functions as the main currency – as in the
roadless parts of the Kola Peninsula. There the rural population swaps reindeer meat
and salmon for fuel with soldiers at the local nuclear submarine base at Gremikha in
the Murmansk Oblast, in complex barter arrangements.[69]

The picture is further complicated by bureaucracy on the part of the regional
and central authorities, and hedging on the part of local towns and settlements.
If there is a surplus, local settlements are not allowed to keep it, and are likely to
have their quota for the next year reduced correspondingly. However, it may be
difficult for the local settlements to predict exactly how much fuel they will need,
depending on how cold the winter is and how well the central heating system
functions. The safest option for local settlements is therefore to request more
than is needed and then discreetly get rid of any surplus at the end of the winter.
Consequently, it can be difficult to distinguish whether settlements need a lot of
fuel because of inefficiency, corruption or hedging.

Another negative factor for expanding the use of renewable energy in the
Northern settlements is that the population is steadily declining. This is partly
due to natural demographic processes, and is partly encouraged by the Russian
authorities.[70] On the other hand, the thousands of remote settlements are not going
to vanish overnight. While some decline, others will inevitably expand along
with the Russian economy, and new ones will have to be established in order to

---

69   Oresheta, Mikhail. 'Umchi menya, olen'. *Polyarnaya Pravda* (28 November
1996), pp. 1–2.

70   A joint programme of the authorities and the World Bank started with a pilot
project to resettle people from the three northern towns of Vorkuta, Norilsk and Susuman
from 1998, and was expanded to resettle 600,000 people from Chukotka, Kamchatka and
Sakha.

Table 2.1   Pros and cons of replacing Northern Freight with renewable energy

Pros	Cons
1. No competition from subsidized gas and electricity via central grids.	1. Parochial interests and corruption give local and central actors involved in Northern Freight incentives to maintain the existing system.
2. No need to connect to a large grid, which is otherwise one of the main obstacles to renewable energy projects in Russia.	2. Potential logistical problems with spare parts and repairs.
3. If efficiency is increased at the same time as renewable energy is increased, considerable synergies can be achieved.	3. The population in remote settlements is declining faster than elsewhere in the country.
4. Renewable energy is already profitable under current conditions in Northern Freight settlements.	

take advantage of the country's vast natural resources. On the Yamal Peninsula, for example, it has been estimated that the population will increase by 25 per cent when Gazprom starts extracting natural gas from the peninsula's deposits sometime from 2011 onwards.[71] A final potential challenge to the use of renewable energy in the Northern settlements is that their remoteness may make it difficult to bring in maintenance staff and spare parts for renewable energy installations in a timely manner. To deal with this challenge, any introduction of renewable energy in remote locations should be done as part of a large-scale state-sponsored programme, in order to ensure economies of scale. If small actors come in and set up a few windmills in remote locations in isolated operations, some of these locations will probably come to lack the spare parts and necessary competence in maintenance, making such renewable energy unreliable.

Despite the obstacles outlined above, the towns and settlements that are the object of the Northern Freight remain one of the most promising prospective markets for renewable energy in Russia today (see Table 2.1). These settlements could emerge as one of the first realistic market niches for the profitable implementation of renewable energy in Russia, while waiting for better framework conditions in the rest of the country. As such, they could function as a testing ground for renewable energy, preparing the country and local and foreign actors for any future expansion in this sector. To succeed, such projects will need the unstinting support of the local and central authorities. There should also be a role to play for foreign actors, in terms of contributing technology, capital and organizational skills.

---

71   Krivitsky and Tsvetinsky 2001, p. 662.

Conclusions

This chapter has analysed some key conditions for the development of renewable energy sources in Russia: the structure of the newly reformed electricity sector and a potential market niche for renewable energy, the Northern Freight system. The necessity of focusing on niche markets is underscored by the fact that the reform of the Russian electricity sector does not address the most important impediment to renewable energy in the country, namely the continued subsidies for domestic gas consumption. Actors seeking to develop renewable energy partnerships with Russia could therefore benefit from focusing on niche markets where competition from cheap gas is avoided. While waiting for the Russian policy-makers to make an energy strategy that significantly improves the conditions for renewables, working in niche markets could serve as a stepping stone to future renewable energy development in Russia on a larger scale.

Chapter 3

The Knowledge-base for Renewable Energy in Russia: Education, Research and Innovation[1]

Jointly, the education, research and innovation systems make up the knowledge-base for the development of renewable energy. An introduction to the Russian scientific-educational system is provided here in order to help actors interested in scientific and education cooperation with Russia understand the country's long history of research and its vast landscape of scientific and education institutions.[2] The amount of research being conducted in Russia declined considerably with the collapse of the Soviet Union, but has started growing again in recent years. This chapter also includes an overview of organizations that fund research in Russia, which may be of use to international research funding organizations wishing to make joint calls; and a ranking of the top education and research institutions in the Russian renewable energy sector. Finally, we examine current Russian policy on innovation, which can be important for the renewable energy sector.

The Soviet heritage

The Soviet Union channelled significant resources into research, and between 1945 and 1975 the science sector grew much faster than other parts of the Soviet economy.[3] Among the strongest areas of Soviet research were nuclear physics, computing, semiconductors, chemistry, materials science, medicine and earth studies.[4] Several of these fields are obviously central to research on renewable energy.

---

1   This chapter was co-authored by Grant Dansie.

2   Aadne Aasland 2007 provides an excellent outline of the science systems in Russia and the Baltic States sponsored by the Nordic Research Board and specifically targeted at a Nordic audience. This section draws extensively on Aasland's outline. Aasland, Aadne. *Development in Research: An Outline of the Science Systems in Russia and the Baltic States.* Oslo: Nordforsk, 2007. http://www.nordforsk.org/_img/nordforsk_pb1_web.pdf [accessed 12 May 2008].

3   Aasland 2007, p. 8.

4   Kovalova, Natalia and Stanislav Zaichenko. 'The Russian System of Higher Education and its Position in the NSI'. Paper presented at *Universidad 2006: The 5th International Congress on Higher Education*, Havana, Cuba, 13–17 February 2006.

Most research, basic research in particular, took place not in universities but in specialized research institutes. The universities, except for a few highly prestigious or specialized ones, were dedicated to teaching. The main centres of research were, and still are, the various institutions under the Russian Academy of Sciences (RAS). Most of these were established before or during the Soviet period, though new ones have also been created afterwards. Currently RAS includes over 430 institutions. They are spread throughout the country and employ an estimated 57,000 researchers and 66,000 auxiliary staff. Aside from the sprawling network of RAS institutions, there are around 1,200 state research institutes. Quite a few of these may be of interest in connection with renewable energy, since they are almost all oriented towards practical technological applications.[5]

There are approximately 685 state institutions of higher education in Russia, of which 48 are universities.[6] In addition many private universities and other institutions of higher education have come into existence since the collapse of the Soviet Union. The most successful private institutions are oriented towards the social sciences, the humanities, journalism, etc., whereas in the natural and technological sciences the state institutions still enjoy top status. Russian basic education comprises only 11 years, but due to the high degree of specialization and pressure to perform in key subjects, the system produces highly qualified students at a young age (as well as a considerable number of losers).

Figure 3.1 gives an overview of some important aspects of the Russian research system. These aspects originate in different periods of modern Russian history, so it makes analytical sense to distinguish between the effects of the Soviet collapse as an event in itself, and the system that has emerged since the collapse.

Brain drain

There has been a double brain drain from Russian science: from research to other sectors of the Russian economy, and from Russia to other countries. Between 1990 and 2002, over half of those employed in Russia's research sector left – an estimated 1,072,000 skilled people.[7] Other negative developments have been no less important, in particular the decline in funding and the lack of opportunities for young scientists. During the past 15 to 20 years it has been far easier for a graduate to get a job in the administration of a university or institute than an academic position. Most grants for young scientists are pittances. As a result, Russian science has many elderly top scientists and good Ph.D. students, but is partly missing the generation of 35- to 50-year-olds. Mending this gap will be difficult or impossible,

5   Aasland 2007, p. 11.
6   Kovalova and Zaichenko 2006, p. 12.
7   UNESCO. *UNESCO Science Report*, Paris, 2005, p. 139.

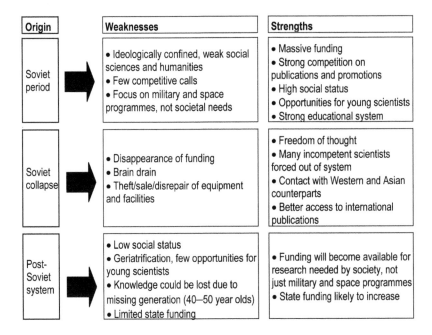

Origin	Weaknesses	Strengths
Soviet period	• Ideologically confined, weak social sciences and humanities • Few competitive calls • Focus on military and space programmes, not societal needs	• Massive funding • Strong competition on publications and promotions • High social status • Opportunities for young scientists • Strong educational system
Soviet collapse	• Disappearance of funding • Brain drain • Theft/sale/disrepair of equipment and facilities	• Freedom of thought • Many incompetent scientists forced out of system • Contact with Western and Asian counterparts • Better access to international publications
Post-Soviet system	• Low social status • Geriatrification, few opportunities for young scientists • Knowledge could be lost due to missing generation (40–50 year olds) • Limited state funding	• Funding will become available for research needed by society, not just military and space programmes • State funding likely to increase

Figure 3.1   Strengths and weaknesses of Russian science, and their origins

because those in that age group cannot simply go back and learn the skills from scratch. Instead it will be necessary to try to build up a new young generation.

Recently the outlook has improved slightly: the central authorities are at least talking more about science; some scientists are returning from abroad or from other sectors of the economy; and large numbers of elderly scientists have retired recently or will do so in the coming years, opening up opportunities for younger scientists. In addition, Russia's largest companies are maturing, and may increase investments in research, given the right incentives. Some examples of this can be seen already: A private–state partnership, the New Energy Projects (NIC-NEP), is a joint venture of the Russian Academy of Sciences and Norilsk Nickel aimed at developing fuel cells, other hydrogen technology and solar power.[8]

Research funding

During the Soviet period, the confines of Marxist-Leninist ideology made it difficult to carry out free research in the social sciences and humanities, because

---

8   See *Novye energeticheskie proekty*. 'Natsionalnaya innovatsionnaya programma'. http://www.nic-nep.ru/ [accessed 23 June 2009].

in order to make a good career it was necessary to avoid overstepping the Soviet ideological boundaries. The natural and technical sciences were more ideologically neutral and could therefore flourish to a greater extent. They were also boosted by massive military spending on research. According to Irina Dezhina, 75 per cent of all research funding passed through the military industrial complex.[9] Almost all of this funding went to the natural and technical sciences.

Research funding was largely distributed through the administrative hierarchy, not through competitive calls. Soviet research was nonetheless highly competitive in prioritized areas, with the competition revolving around empirical results and publications.

The economic situation in the 1990s made research a low priority for both politicians and the public. Government research funding was drastically reduced, and the decline in industrial production also led to a decrease in research commissions from industry. Out of necessity, many academic institutions entered into commercial activities; some of them were even compelled to sell off equipment and cut back on staff. The economic difficulties also led to a drastic reduction in investments in applied and basic research without immediate commercial value.[10]

In 2004, the former Ministry of Education and parts of the Ministry of Industry, Science and Technology were merged into a joint Ministry of Education and Science, to be responsible for legislation and policy on research, education, innovation and intellectual property rights.[11] Although the ministry has been criticized for failing to secure sufficient resources for basic research, much of the decrease in funding for basic research occurred before the ministry was created, and this decrease reflects the decline in overall funding for research. Such criticism could therefore rather be seen as reflecting the stronger emphasis that many researchers and research policy actors would like to see on basic research, a continuation of trends from the Soviet period. In any case, research funding remains very low in Russia, both compared to the Nordic countries and the EU. Russia is even outclassed by Norway, the Nordic laggard.

In recent years the research and development share of Russian GDP has increased, and the disbursement of state funding is now more timely and predictable than during the difficult 1990s. The state is still the primary source of research

---

9   Dezhina, Irina (2002) 'American Science Foundations in Russia as Driving Forces of International Transfer in Knowledge and Professional Skills'. Paper presented at the conference *Transforming Civil Society, Citizenship and Governance: The Third Sector in an Era of Global (Dis)Order*, University of Cape Town, 7–10 July 2002. http://www.istr. org/conferences/capetown/volume/dezhina.pdf [accessed 5 March 2008].

10   Aasland 2007, p. 14.

11   Aasland 2007, p. 10.

funding in Russia, with industry accounting for approximately one-third of the funding.[12] However, state funding is taking a different shape now compared to the Soviet period. There is a movement towards making government research and development funding more transparent, target-oriented and efficient, and more competition is being introduced.

One of the most highly respected institutions for research funding in Russia, the Foundation for Basic Research, is widely known for funding large-scale projects through transparent and highly competitive calls. As in most countries, however, far more research funding is channelled directly through the state universities and research institutes than through such competitive research funding institutions.[13]

An overview of some key Russian funding institutions for research that are of relevance for renewable energy development and that could be relevant for international actors interested in co-financing is provided in the appendices.

Top Russian institutions

In trying to pinpoint which Russian institutions should be targeted for cooperation on renewable energy, we have compiled a ranking of the top institutions in research and education on renewable energy in Russia. Table 3.1 presents a ranking based on the opinions of 12 experts, and also draws on three older ratings of educational institutions.

This ranking is intended to serve as a rough guide for those searching for Russian partners for international projects. As mentioned, the country has over 430 research institutes just within the Academy of Sciences, and another 1,200 state institutes outside the Academy.[14] In addition there are a good many private institutes. Foreign actors may not be well acquainted with the many Russian research and education institutions, and we therefore hope that this ranking can serve as a basic who's-who for those trying to orient themselves in the Russian renewables research landscape.

Due to the limited data basis, the ranking should be considered tentative and preliminary rather than final. It nonetheless gives an interesting glimpse of what may be the top institutions when it comes to renewable energy in Russia. Explanations of the data and methodology behind the ranking and details about the

---

12 Khutorova, Natalia. 'National research funding in Russia', Moscow State Forest University, 2008. http://www.ftpc5.si/files/FTP%20PDF%20Presentations/Tuesday%20pa rallel%20WoodWisdom/WW%20Slovenia%20Russia.pdf [accessed 21 August 2008].
13 Aasland 2007, p. 11.
14 Aasland 2007, p. 11.

Table 3.1   Ranking of top Russian institutions

Rank group	Title	Score
1	Bauman Moscow State Technical University (MSTU)	
	Moscow Power Engineering Institute (MPEI)	100–150
	St. Petersburg State Polytechnic University	
2	All-Russian Institute for the Electrification of Agriculture (VIESH)	
	Novosibirsk State Technical University (NSTU)	
	St. Petersburg State Mining Institute	50–100
	United Institute of High Temperatures of the RAS	
	Urals State University	
3	RusHydro Scientific Research Institute of Energy Construction	
	Goryachkin Moscow State Agro-Engineering University	
	Ioffe Physico-Technical Institute of the Russian Academy of Sciences	
	Ivanovo State Power University	
	Krzhizhanovskiy Power Engineering Institute (ENIN)	20–50
	Lomonosov Moscow State University (MGU)	
	Moscow Aviation Institute (MAI)	
	Moscow State University of Ecological Engineering	
	Tomsk Polytechnical University (TPU)	
4	Boreskov Institute of Catalysis SB RAS Novosibirsk	
	Buryat State Agricultural Academy	
	Central Aerodynamics Institute	
	Dagestan State University	
	Energy Strategy Institute	
	International Science and Technology Centre (ISTC)	1–20
	International University of Nature, Society and Humanity 'Dubna'	
	Kazan State Energy University	
	Kostyakov All-Russian Institute of Hydrology and Irrigation	
	Mari State Technical University	
	Moscow State University of Railway Engineering	

top institutions are presented immediately after the ranking table. An alphabetical list of all of the ranked institutions (including their Russian names and web addresses) can be found in the appendices.

Ranking

The ranking list is based on a survey of the opinions of 12 experts who were asked to rank as many institutions as they could, according to their prominence in (1) research on renewable energy and (2) education on renewable energy. The experts were asked to take into account both the excellence of the institution and its specific focus on renewable energy. Thus institutions which are generally very prestigious but are not seen as focusing specifically on renewable energy could score lower than those that are both excellent *and* focus specifically on renewable energy. Conversely, an institution that works exclusively on renewable energy but has no merit could be ranked relatively low. In order to secure free and unfettered opinions, the 12 experts were guaranteed anonymity, except that it was understood that they would be among the more than 100 interviewees listed in the appendices.

In basing the ranking on the perceptions of a selection of experts, we were inspired by among other things the methodology applied in some of the most widely quoted corruption rankings.[15] We handpicked the 12 experts while carrying out general interviews for this book. When we came across a respondent who we thought might have a good overview and in-depth understanding of renewable energy in Russia, we asked him or her to help us rank institutions. The input of interviewees who did not appear to have sufficient knowledge was discarded. Taking the ranking seriously therefore depends both on one's trust in the judgements of the experts, and one's trust in our judgements in selecting the experts and communicating with them.

Six of the experts were men, six were women. All twelve were based in Russia: eight in Moscow, two in St. Petersburg and two in other parts of the country. There is therefore a risk of a Moscow bias in their judgements. On the other hand, much of Russia's academic elite is gathered in Moscow, so our selection is not wildly disproportionate. Among the 12 experts, the largest group were academics, but there were also representatives of NGOs, government and business. The experts represented a broad age-span, with roughly a third in their 30s, a third in their 40s or 50s and a third in their 60s or 70s. Apart from its small size, the expert panel therefore provides a reasonable approximation to the spread of backgrounds, as recommended by Dannemand Andersen and Holst Jørgensen.[16]

All expert rankings were carried out through face-to-face interviews. As pointed out by Dannemand Andersen and Holst Jørgensen, this enables those gathering information to guide the interviewees and explain questions in greater detail when needed.[17] Our impression was that the respondents approached the task with great seriousness, and did not attempt to promote their own institution or those of their contacts. Some were very reluctant to carry out the task, explaining that it was difficult for them to give such opinions offhand. Some felt competent to rank only

---

15   See for example the Transparency International Corruption Perceptions Index (Graf Lambsdorff 2007), although this has grown to a much larger and more complex survey in recent years. Graf Lambsdorff, Johann. *The Methodology of the Corruption Perceptions Index.* Berlin: Transparency International, 2007.

16   Dannemand Andersen, Per and Birte Holst Jørgensen. *Grundnotat om metoder indenfor teknologisk fremsyn.* Risø: Forskningscenter Risø, 2001, p. 7.

17   Dannemand and Jørgensen 2001, p. 7.

research institutions, others only educational institutions. Some would rank only two or three institutions, whereas others found it easy to reel off long prioritized lists of both research and educational institutions with ten or more institutions on both lists.

In addition to the input from our 12 selected experts, three pre-existing surveys of Russian educational institutions were used:

1. Top educational institutions in energy studies[18]
2. Potanin Fund ranking of top educational institutions[19]
3. Ranking of institutions that produce most rectors of institutions of higher education and members of the Russian Academy of Sciences.[20]

Each of these surveys was given the same weight as one expert opinion. We considered the option of instead multiplying the three older rankings by a factor of three, to give them triple weight. In that case, a top rank of an institution in one of the other rankings would have given 45 points in ours (15 x 3), a second-place rank would have given 42 points (14 x 3), etc. This could have been justified by arguing that (1) these are large-scale independent rankings which should carry more weight than any individual expert consulted by us, and (2) the experts would in any case exercise considerable power by getting to make the initial selection of institutions, with the other rankings used only to modify the subsequent ranking among these institutions.

This solution was, however, not chosen, since the three older rankings that are used all focus on educational institutions, and this would have given education too much weight compared with research. After all the experts had been consulted, each rank was turned into a score. A top ranking would give 15 points, a second ranking would give 14 points, etc. (No experts ranked more than 15 institutions.) Those experts who felt competent to rank more institutions thus got to dish out more points than those who felt they could rank only a few institutions. Thus our ranking is also based on a self-selected weighting of the most competent experts.

It is our opinion that the ranking provide a pertinent account of what may be the top institutions in renewable energy in Russia. We hope that the ranking can serve as a rough guide for those searching for Russian partners for joint projects. The ranking does not, however, purport to be objective or final. In order to identify

---

18   Ucheba.ru. 'Reyting vuzov po kriteriyu usloviya obucheniya v vuze'. N.d. http://www.ucheba.ru/vuz-rating/2856.html [accessed 23 June 2009].

19   Vladimir Potanin. 'Fond Potanina sostavil reyting vedushchikh rossiyskikh vuzov'. N.d. http://www.ucheba.ru/vuz-rating/1922.html [accessed 23 June 2009].

20   Svetlana Danilova. 'Akademicheskaya elita, Reyiting vuzov 2007'. N.d. http://www.ucheba.ru/vuz-rating/1922.html [accessed 23 June 2009].

and create an accurate hierarchy of the top institutions on renewable energy in Russia, a far larger survey would be needed, and it might also be desirable to include citation indexes as one component. This would require far more resources, as it would involve identifying all of the key researchers on renewable energy in each institution, and then checking their citation indexes.

Basing such a ranking on the views of only 12 experts and three pre-existing rankings thus limits the validity of our findings, which should therefore be considered preliminary rather than final. This is also why we have not ranked institutions one at a time, but bunched them into four main groups. Within each group the institutions are listed alphabetically.

Details of research and education institutions ranked

In this section, further information is provided about the eight institutions in the two top groups in the ranking list.

Bauman Moscow State Technical University (MSTU)

MSTU has a history dating back to the 1830s, when it was founded with the aim of training skilled labour with a solid theoretical background in order to improve and spread skills in various trades all over Russia. The history of MSTU is like a mirror of Russian and Soviet history, as it has followed several industrial trends closely. From the late imperial period, the school specialized in training engineers, and in the 1930s it branched out into several institutions of higher learning which now focus on renewable energy technologies, among others the Moscow Power Engineering Institute (see below). From 1938, MSTU had a distinct defence profile, and in 1948 a department of rocket science was opened. Now in the post-Soviet period it emphasizes good relations with industry and high professional and moral standards for both students and staff. The main strength of MSTU lies in education, and it boasts a long-standing tradition of Russian training in engineering, but it is also strong in research.

MSTU aims to provide its students with the technological skills needed to meet the needs of the Russian economy, and considers energy and energy efficiency as one such area. Compared to other Russian research institutions, it can be said to be at the forefront of intellectual property issues, and boasts a centre for the protection of intellectual property, aimed at helping university staff to obtain patent rights.

Moscow Power Engineering Institute (MPEI)

During the Soviet period, MPEI was known as a global centre of power engineering, so it is one of the most internationalized institutions included in our ranking. It received students from communist and developing countries all over the world, often from powerful families. As a result it currently has alumni in prominent positions, including ministers, heads of energy companies, etc. in numerous countries. China, for instance, has its own alumni organization for MPEI alumni. Its researchers were also involved in projects on several continents, for example building dams. MPEI's strength is in education, although it is also engaged in research. MPEI is also a key institution in the Russian context, and its former students dominate the state organs and power companies. MPEI formerly had a Faculty (*kafedra*) of Hydroenergy, which was renamed the Faculty of Non-Traditional Sources of Energy in the 1980s. In most of the post-Soviet period, the official head of the faculty has been a manager of RAO UES. On a day-to-day basis the faculty is run by its most prominent professor, Vissarionov.

St. Petersburg State Polytechnical University (SPGPU)

SPGPU was originally founded as the Saint Petersburg Polytechnic Institute in 1899. It is a full-scale technical university that covers most technical and technological topics. Its main strength is in education, although it is also engaged in research. A 'Faculty for the Utilization of Hydro-Energy' was set up in 1921, renamed the 'Department for Renewable Sources and Hydropower' in 1986. It is placed under the Faculty of Civil Engineering, which has 40 professors and doctors of science and 840 students. The department deals with an extensive range of renewable energy forms, including solar, wind, bioenergy and small and big hydro. The SPGPU has a training programme in innovation which aims at providing students with the skills needed to produce innovations in prioritized technical and research areas. This programme has energy efficiency technology as one of its four main areas.

All-Russian Institute for the Electrification of Agriculture (VIESH)

VIESH may have an obscure-sounding name and an even more obscure-looking website, but it is clearly one of the main institutions involved in research and education related to renewable energy in Russia. In fact, it is precisely the connection to agriculture that helped secure VIESH its central position in Russia's renewable energy sector. When the Soviet authorities decided to scrap most research into renewable energy and focus on nuclear power and natural gas instead, they still had to admit that the more remote farms and settlements in the vast Soviet territory would be too far distant to connect to the electricity and gas grids. VIESH has therefore been able to continue its work on renewable energy without interruption, from its inception in the early Soviet period until today. The institute covers a broad range of renewable energy, including photovoltaics and solar collectors. It is equally strong in education and research, and has since 1997 hosted the UNESCO Russian chair in Renewable Energy and Rural Electrification, Professor Dmitry Strebkov.

Novosibirsk State Technical University (NSTU)

The NSTU was founded during the Second World War to train specialists for the industry that moved east of the Urals to the Novosibirsk region to avoid the fighting. With approximately 25,000 students, it is one of the largest universities in Siberia. The Energy Faculty of the NSTU was founded in 1962, and has been dealing with hydropower and energy efficiency since its inception. It carries out a broad range of research projects and trains students in most energy-related fields. The NSTU also hosts the Novosibirsk Regional Power Savings Centre.

St. Petersburg State Mining Institute

This institute, founded in 1773, is one of Russia's oldest technical educational institutions. In addition to its focus on mining, geology and hydrocarbon resources, it carries out general energy research and is among the leading institutions in geothermal energy.

United Institute of High Temperatures of the Russian Academy of Sciences

This is one of the world's leading institutes of physics and other hard sciences. The reason it is not included in the top group is therefore not lack of excellence, but rather that renewable energy is a relatively small part of its activities, though still significant.

Urals State University

The university was founded in Ekaterinburg in 1920. It has approximately 10,000 students, and is especially strong in physics and applied mathematics.

Russian innovation policy

The expansion of renewable energy is dependent on appropriate innovation policies to facilitate the investment of capital and the realization of good ideas. If international actors are to cooperate with their Russian counterparts, it may be helpful to understand Russian policy on innovation, since this is part of the context within which they will have to operate together.

With significant empirical evidence indicating the importance of innovation for long-term economic growth, this would appear to be a key focus area if Russia is to achieve the standing as a significant global economic player that the Kremlin yearns for.[21] Although the Russian economy is currently experiencing high levels of growth, this is largely dependent on high commodity prices. Focusing on research and innovation would help foster new industries, increase productivity and diversify the economy. Renewable energy is attractive in this respect. Add to this the country's extensive natural resources, and the field of renewable energy would appear ripe for investment.

But what do we mean by innovation? In a business sense 'the goal of innovation is to create business value by taking ideas from mind to market'.[22] Mckeown defines innovation more broadly as referring to both radical and incremental changes in thinking, in things, in processes or in services.[23] In terms of renewable energy, 'innovation' is used to refer to developing technologies, services and equipment that will lead to increased productivity and technological advancement in the renewable energy sector. Unfortunately in the Russian context, innovation has often failed to lead to market or business value, due to problems in commercialization.

Russia has considerable innovation potential, benefiting from an extensive science base and well-developed educational system in the areas of science and technology. And yet, Russian innovation indicators remain disappointing.[24] There has long been a disparity between the amount of public resources devoted to innovation and the results achieved. A 2006 OECD report by Gianella and Tompson on the state of innovation in Russia highlights this as one of two key challenges facing innovation in Russia, the other being to stimulate private-sector participation in research and development – a challenge that many West European countries are also struggling with.[25]

History of Russian research and development

The beginnings of modern Russian research can be traced back to 1724, when Peter the Great founded the Russian Academy of Sciences, which became world renowned for its excellence in science. Merkina highlights how the practice of carrying out most research in institutes of the Academy of Sciences, at leading

---

21   Donselaar *et al.* 2004; Keller 2004, both quoted in Gianella, Christian and William Tompson. *Stimulating Innovation in Russia: The Role of Institutions and Policies.* Working Paper. Paris: OECD, 2006, p. 5.

22   Smith, Howard. *What Innovation Is: How Companies Develop Operating Systems for Innovation.* CSC White Paper for the European Office of Technology and Innovation, 2007.

23   Mckeown, Max. *The Truth about Innovation.* Harlow: Pearson Education Limited, 2008.

24   Gianella and Tompson 2007, p. 5.

25   Gianella and Tompson 2007.

universities and military laboratories goes back to pre-revolutionary times.[26] Only a small number of industrial firms had internal research departments. Thus it was these key institutions that provided the foundation for the Russian research and development system after the Revolution of 1917.[27]

During the Soviet period, each branch ministry had its own research institute(s) which carried out branch-wide research and development, instead of individual enterprises carrying out research.[28] Thus when the Soviet Union collapsed, there was relatively little enterprise-specific research underway, which was obviously closely related to the subsequent poor levels of privately-funded R&D and the lack of commercialization of innovative activities. This also highlights the dearth of innovation, as enterprises had no incentives for innovating within the framework of a planned economy. Nor had they any obligation to innovate, as research was carried out by the branch ministry. Watkins highlights how this monopoly of research by the branch ministries resulted in the quality of the output being generally below world standards.[29] This could be attributed to a lack of competitiveness, and a lack of incentive to innovate. The Soviet science and technology system could be seen as a set of parallel worlds, with few if any linkages or communication and feedback channels between silos. This bureaucratic stratification was further enhanced by geographical stratification, with more than 50 closed science cities. For reasons of security, these were either situated in very isolated locations, or in gated, restricted communities adjacent to civilian cities.[30]

Current role of the state

Russian research and development is still primarily financed by the state. Approximately 60 per cent is publicly funded – a significant contrast with the OECD countries. Around 80 per cent of Russia's research personnel work in the public sphere, part of approximately 2,900 publicly owned research institutions.[31]

Funding is divided between the Russian Academy of Sciences, the applied research centre and the higher education sector. The Academy of Sciences has usually been allocated around 40 per cent of budget funding for science, with the

26   Merkina, Natalia. 'Innovation and Regional Development in Russia'. MSc thesis University of Oslo, Environmental and Development Economics, 2004, p. 16.
27   Merkina 2004, p. 16.
28   Watkins, Alfred. *From Knowledge to Wealth: Transforming Russian Science and Technology for a Modern Knowledge Economy*. World Bank Policy Research Working Paper no. 2974, 2003, p. 9.
29   Watkins 2003, p. 9.
30   Watkins 2003, pp. 7–9.
31   Gianella and Tompson 2007, pp. 7–22.

Russian Academy of Medical Sciences receiving 6 per cent and higher educational institutions around 5 per cent.[32] Most of this funding is provided with few strings attached, and is allocated on a cost basis according to employee levels and fixed assets. Gianella and Tompson hold that such institution-based funding creates few incentives to increase efficiency, productivity or innovation and tends to protect incumbents.[33] They argue for a significant shift towards competitive allocation and project-based funding.

Private sector

Innovation is rather low in the private sector, with business expenditure on technological spending corresponding to only 1.5 per cent of industrial sales in 2004, and a mere 3.3 per cent for firms engaged in innovative activities.[34] The private sector has focused on imitation rather than research-based innovation, and over 50 per cent of business expenditure on technological innovation has aimed at improving production processes rather than developing new products.[35]

The Russian information and communications technology (ICT) sector provides an effective example of the constraints facing innovation. This sector has shown significant growth in recent years, with ICT spending increasing 27.8 per cent in 2005, accounting for 4.7 per cent of GDP.[36] There is still massive potential for growth, but various structural constraints hinder this growth – lack of IT specialists, lack of labour mobility, poor infrastructure and low levels of research and development.[37]

Commercialization

Poor communication between the public and private sectors has affected the levels of commercialization. With most research and innovation being state-funded, researchers in the public sphere generally have little reason to think about commercial applications of their work.[38] Moreover, the state has retained the rights to the results of publicly financed research – leading to debate between various groups within the government and research establishment about intellectual

---

32   Gianella and Tompson 2007, p. 20.
33   Gianella and Tompson 2007, p. 20.
34   Gianella and Tompson 2007, p. 7.
35   Gianella and Tompson 2007, p. 7.
36   Gianella and Tompson 2007, p. 11.
37   Gianella and Tompson 2007, p. 11.
38   Gianella and Tompson 2007, p. 11.

property rights.[39] A law was introduced in 2005 allowing this issue to be discussed between the financing bodies and organizations carrying out research and development, in order to attract investors and promote commercialization. Many Western governments also hold rights to the research generated in their state-funded institutions, but these countries have developed incentive mechanisms to encourage researchers to commercialize their research. This contrasts with the Russian situation, where wholesale state ownership of results has created a disincentive for research institutions to engage in innovation.

A second major constraint on commercialization is Russia's weak intellectual property rights (IPR) framework. A World Economic Forum survey on the perceptions of levels of IPR protection among Russian businesspeople led to Russia being ranked 105th out of 117 in terms of IPR protection. Additionally, in a survey by the Interdepartmental Analytical Centre, half the respondents cited weaknesses in the IPR regime as a major constraint to the commercialization of research.[40]

Educational backgrounds can also have an effect on commercialization. Merkina highlights how the Russian industrial managerial class is still heavily dominated by scientists and engineers, who tend to focus on technological attainment rather than commercialization.[41] International actors aiming to carry out commercialization in cooperation with Russian partners must therefore explicitly bring out both the benefits to be achieved from commercialization and the opportunities available.

Watkins highlights how the lack of domestic demand has also influenced the pattern of innovation in Russia:

> Science-intensive enterprises and research institutes typically find that there is relatively little demand for their goods and services inside Russia. Instead, their most lucrative markets seem to be outside Russia, either in other emerging markets or occasionally in Western Europe or the US. Thus, while most countries are integrating their science and technology sector with a vibrant, globally competitive domestic enterprise sector, Russia would appear to be developing two independent systems – an enterprise sector that occasionally finds the financial resources to purchase technology and knowledge-intensive equipment from abroad, and a science and technology sector that occasionally manages to sell Russian technology and knowledge-intensive equipment abroad.[42]

---

39   Watkins 2003, p. 23.
40   Quoted in Gianella and Tompson 2007, pp. 19–20.
41   Merkina 2004, p. iii.
42   Watkins 2003, p. 3.

Government reform plans

The Russian government appears to be aware of the need to reform its innovation policy, and has launched a strategy for the development of science and innovation to 2015. This includes both institutional reform and targeted initiatives. We start by outlining the latter.

The most significant initiative is the creation of four technical innovation zones, called Special Economic Zones, which have a range of tax and customs incentives. Next, the plan is to develop a network of eight technoparks throughout the country, with business incubators, technology transfer centres and other innovation infrastructure.[43] The regions will be responsible for creating the parks, which will not enjoy tax or customs preferences, but will receive financial support from the state.

Another measure will be the creation of 'science towns', which in turn will be able to create technoparks and innovation 'business incubators' on their territory. These towns will receive federal funding in order to develop their science base.[44] It is unclear whether these towns will be new or will in fact be the old *akademgorodki* of the Soviet era.

Various venture capital funds have also been created under the auspices of different ministries – Economic Development and Trade, Industry, Science and Technology and Information Technologies and Communications – seeking to promote innovation in a range of forms. In their 2006 OECD report on innovation in Russia, Gianella and Tompson note the difficulties in measuring the success of innovation policies:

> It is an experimental science, and the government should proceed in that spirit, viewing ... targeted interventions as experiments requiring rigorous evaluation and review at regular intervals, as well a willingness to drop initiatives that fail to produce results.[45]

In terms of institutional reform, the government plans to change the organizational structure of Russia's research institutes, transforming them into 'autonomous institutions'. This move is intended to give them both greater financial freedom and greater responsibility.[46] The Ministry of Education and Science hopes to change the ratio of institutional funding to project funding from 80/20 to 50/50 in order to spur competition and innovation.

43　Gianella and Tompson 2007, p. 25.
44　Gianella and Tompson 2007, p. 25.
45　Gianella and Tompson 2007, p. 30.
46　Gianella and Tompson 2007, pp. 21–2.

Ultimately these new innovation policies will be important, but the success of innovation in Russia may depend on broader institutional and economic factors such as effective public administration reform, reduced corruption and the creation of a healthy business environment. Russia needs more private investment in innovation – yet, without predictability and stability, companies will be reluctant to invest. At the same there is a need to generate competition for funding within the state's institutional framework, whilst strengthening the legal underpinnings of new legislation and protecting intellectual property rights.

Chapter 4
Russia's Solar Power Sector

In our work on this book, we have identified two scientific fields in which Russia excels and which are relevant for the development of renewable energy: solar power and hydrogen. Here we provide an overview of the most scientifically mature and financially largest of these two sectors: solar power. Our focus is on the historical development of this sector during the Soviet and post-Soviet periods, with an overview of the current actors and policy and markets. In order to understand the constraints and possibilities in this sector, it is also necessary to explain some of the basic technological aspects of solar power.

Solar power can be generated in two ways: (1) photovoltaics (the photoelectric transformation of solar radiation into electricity) and (2) solar thermal power (the concentration of heat from sun rays in solar collectors in order to heat water or other substances). Russian scientists and manufacturers are engaged in both fields, but the cutting edge in this area of Russian science has been photovoltaics, which is the topic of this chapter.

The Soviet Union was one of the world leaders, at times *the* world leader, in basic research into photovoltaics. Soviet photovoltaics had two main applications:

1. The Soviet space programmes, where solar panels were used to supply satellites. The first solar-powered satellite was launched in 1958.
2. Electricity generation for remote installations on the ground, and potentially for larger settlements, mainly in the southern parts of the USSR.

This scientific tradition in photovoltaics was undermined by the demise of the Soviet Union, for three main reasons. Like most areas of basic research, it suffered from a sudden decline in funding; the development of new applications for space programmes was especially hard hit by the end of the Soviet Union; moreover some of the most promising areas for the use of photovoltaics were in the southern Soviet republics, which became independent after the fall of the Soviet Union. Uzbekistan was one of the major centres for scientific developments in photovoltaics. It was home to the main Soviet scientific journal on the topic, *Geliotekhnika*, published by the Academy of Sciences of Uzbekistan. Any young academic who managed to get an article published in *Geliotekhnika* had ensured his or her career. With the demise of the Soviet Union, the links between research groups in Uzbekistan and Russia were severed or weakened, leaving each individual milieu weakened.

Since the year 2000, however, there has been a resurgence in the Russian photovoltaics sector. This is due mainly to the increasing demand and prices for photovoltaic cells and modules on world markets, but also to the nascent interest in renewable energy in Russia under the influence of global trends.

Some actors have held that solar radiation cannot be a realistic source of energy in Russia, since large parts of the country lie at high latitudes and have a cold climate. This view partly builds on mistaken assumptions. Firstly, much of the country, especially central and southern Siberia, has an exceptionally continental climate with frequent high-pressure areas and many cloudless days. Secondly, although much of Russia does have a cold climate, photovoltaic solar panels do not generate electricity from heat, but directly from the sun's radiation. The key variable therefore is not temperature but rather access to direct sunlight. In fact, heat decreases the efficiency of photovoltaic panels. So in the cold climate of Russia, solar panels can achieve greater efficiency than in locations with the same amount of sunshine but higher temperatures. Nonetheless, due to economic and political factors, the market for solar power in Russia remains very limited for the time being. The main driver for the sector's nascent rebound is exports.

In addition to the general barriers to renewable energy in Russia discussed in Chapter 2 (subsidies for hydrocarbons and nuclear power, lack of subsidies for renewables, difficulties connecting to grids), solar power faces some specific challenges in the Russian context:

- The most populous parts of the country are not those with the most sunshine (see Figures 4.1 and 4.2). Exceptions to this rule are the Far-Eastern city of Vladivostok and densely populated and rapidly growing Krasnodar Krai in the country's southwestern Caucasus region.
- The population is uninformed about the potential of solar power in Russia. There is a lack of demonstration sites and government-sponsored information.

Solar radiation in Russia

The two maps opposite show that different sources give differing figures as to the distribution of sunlight hours in the Russian Federation. We believe that the second map is the more accurate of the two, but both show that there is sufficient potential for the development of solar power installations in large areas of the country. Over 60 per cent of Russia's territory, including some parts in the north of the country, have adequate sunlight for solar panels (3.5 to 4.5 Kwh/m^2 per day). The sunniest regions are in the Far East and southern Siberia, with 4.5 to 5 Kwh/m^2 per day. Much of Siberia, including Sakha, as well as the Northern Caucasus and Sochi receive 4 to 4.5 Kwh/m^2 per day.

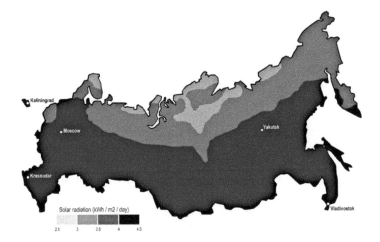

Figure 4.1   Solar power potential in the Russian Federation 1. Contrast with map below[1]

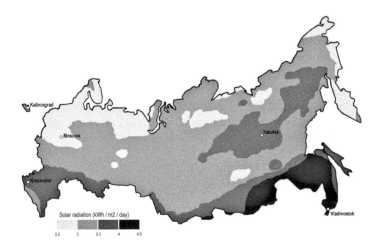

Figure 4.2   Solar power potential in the Russian Federation 2. Contrast with map above[2]

1   Source of data used to compile figure: Baranovskiy, Sergey and Aleksandr Chumakov. 'Alternative Energy in Russia: Problems and Perspectives'. *Alternativnaya Energetika*, vol. 7, no. 1 (2008): 2–6.

2   Source of data used to compile figure: Popel, Oleg S. 'Tekhnologii i sfery effektivnogo energeticheskogo ispolzovaniya vozobnovlyaemykh istochnikov energii v regionakh Rossii'. Presentation at the seminar *Itogi realizatsii proektov v ramkakh prioritetnogo napravleniya 'Energetika i energosberezhenie' federalnoy tselevoy nauchno-tekhnicheskoy programmy v 2007 g.*, Moscow, Russia, 7 December 2007.

The basics of silicon production

Several different materials can be used to build photovoltaic equipment for electricity production. However, for commercial equipment, semiconductor silicon is invariably used. The production chain for silicon-based solar power equipment is complex and requires considerable inputs of energy. It can be divided into two main parts:

1. The production of silicon and silicon plates, which is the raw material for the preparation of most solar modules, which are subsequently assembled into solar batteries.
2. The production of solar elements, and assembly into modules and batteries.

Here we look at this series of operations in terms of the 'Siemens Process', meaning the process of exposing high-purity silicon rods to a silicon compound, trichlorosilane, at a temperature of 1150°C in order to enlarge them, since this is what has generally been applied both in the Soviet Union and Russia.

The basic material for the production of semiconductor silicon for the electronics and photovoltaic industries is metallic silicon (MeGSi), which in Russia is generally imported and is consumed in large quantities by the country's steel industry. (In the Soviet Union, the production of semiconductor silicon was never linked with the production of metallic silicon, as it has been elsewhere.) The main problem now for the development of the photovoltaic industry in Russia, as it is for the industry worldwide, is the lack of polycrystalline silicon, which is the basic input material for the fabrication of monocrystalline silicon for the electronics sector and for the fabrication of both types of semiconductor silicon (monocrystaline and multicrystalline) for solar cell production.

The history of Russian photovoltaics

In the Soviet Union, the first research on the creation of photovoltaic transformers was carried out at the Physical-Technical Institute of the Academy of Sciences in Leningrad, which was led by the famous academician Abram Ioffe. Soviet scientists first managed to generate electricity through a photo effect in the 1920s, although the initial efficiency was only 1 per cent. Ioffe was a great enthusiast of photovoltaics. When two of his young researchers in 1930 managed to create a new photo-element with an efficiency of 1 per cent, he proposed a large-scale state programme to install such elements on rooftops.

Subsequently the centre for the development of photovoltaic transformers and their practical application moved to the All-Union Scientific Research Institute on sources of electricity (VNIIT) in Moscow, which was responsible for the electricity supply of all satellites. (After the dissolution of the Soviet Union,

VNIIT was renamed Kvant and became a stock company.[3]) One of the key players at VNIIT was academician Nikolay Lidorenko, whose name is closely tied to the development of Soviet photovoltaics and the trajectory of VNIIT.

Already in 1958, a solar panel was installed on the third Soviet satellite. (In the USA there were parallel developments and similar equipment was installed on the satellite Avantgarde the same year.) See Table 4.1 for an overview of the milestones in the evolution of Russian nuclear power.

In 1970, Alferov and his colleagues created a solar cell based on gallium arsenide (GaAs) which was more expensive but also far more efficient than silicon-based systems. By the mid-1970s, Soviet silicon-based solar cells, similar to silicon-based solar cells in the rest of the world, achieved almost 10 per cent efficiency, but then remained at that level for the next two decades. For the space applications of the time that was sufficient, but for terrestrial use it meant that solar power remained highly expensive and unattractive for further investment compared to other energy forms. This was the case around the world, but above all in a major petroleum-producing country like the USSR. Much research on new solar power technology came to a halt as financing faded away. In other parts of the world, the 1970s oil crises gave a counter-impulse for accelerated research on solar power, and research to increase the efficiency of silicon-based solar cells continued to some extent.

The USSR, however, was not dependent on oil imports and therefore did not suffer from the 1970s oil shocks, and therefore had even less incentive to continue intensive work on solar power. Only much later, after the collapse of the Soviet Union, did some physicists resume research in this area, and still without significant state support. It is thanks to these individuals, who were trained in the Soviet tradition and who continued fuelled by their personal enthusiasm, that Russia today has a cluster of modern companies producing solar cells and modules. Despite the lack of government support, they were able to raise the efficiency of solar cells to 15 per cent by the mid-1990s, and to almost 20 per cent by the year 2000. In the early 1990s, Nobel Laureate Alferov of the Ioffe Institute argued at an assembly of the Soviet Academy of Sciences that if 15 per cent of the funding then being spent on nuclear energy were spent on renewable energy, there would be no need for nuclear plants. However, such views failed to evoke greater engagement from the state, which at this point was disintegrating anyway.

---

3   The company Kvant (former VNIIT) continues to carry out active research on solar energy for both cosmic and earth-based applications. Already in the 1970s systems were created and installed in several parts of the Soviet Union for purposes such as water pumps, navigation systems for shipping lanes, telecommunications and so on. During the past two years, Kvant has established a 5 MW per year production line for silicon-based solar cells using Russian-produced equipment. This plant can be expanded to a capacity of 20 MW or more per year. The solar elements have an efficiency of 16 per cent.

Table 4.1   Main milestones in the evolution of Russian solar power[4]

1958	Solar-powered satellite
1967	Multi-junction matrix cell technology with voltage 10–100 V/cm²
1970	Bifacial solar cell · Ion implantation cell technology
1975	GaAlAs-GaAs solar cell for Moon and Venus missions · 1 m², 32 kV solar array for ion plasma space engine
1980	GaAlAs-GaAs multi-junction solar cell
1981	Compound, golographic, prism concentrator technology
1983	Grid connected 10 kW parabolic trough solar power plant
1985	Laser PV energy conversion (electric power 3.6 kW/cm², 36 per cent)
1987	Purification of metallurgical-grade silicon
1993	30 per cent cascade multi-junction GaAlAs-GaGeAs cell
2000	Stationary parabolic trough concentrator with concentration ratio 3.5–14
2001	Resonance single-wire electric power transmission line 20 kW, 10 kV
2002	Chlorine-free technology for silicon production
2003	Vacuum glazing with thermal transmittance 0.5–2.27 W/m²K
2004	Global Solar Power System with 24-hour year-round generation

Non-silicon solar panels

The global deficit of polycrystalline silicon has spurred Russian actors to carry out research into the production of solar power modules based on the highly efficient Gallium-Arsenide (GaAs) semiconductor, which has an efficiency level higher than 30 per cent. Currently the company Kvant is developing solar elements of this type for cosmic applications, and is preparing to start production. Kvant and the Ioffe Institute have begun a research programme on these solar modules, ensuring that there will be further research in this direction.

Work to combine photovoltaic solar power modules using concentrators in the form of Fresnel lenses has great potential, for several reasons.[5] Firstly, the use of concentrators enables much higher efficiencies. Secondly, when concentrators are used, far fewer solar cells are required, which reduces the need for expensive semiconductor materials. At the All-Russian Institute for the Electrification of Agriculture (VIESH), Arbuzov and Evdokimov have shown that it is theoretically possible to make solar cells with an efficiency of 93 per cent using this method. Researchers from the Ioffe Institute have also arrived at similar efficiencies

---

4   Source of data used to compile table: Strebkov, Dmitry S. 'Large-Scale Renewable Energy Technologies', Presentation at the *3rd International Conference on Materials Science and Condensed Matter Physics*, Chisinau, Moldova, 3–6 October 2006.

5   Fresnel lenses are traditionally used in lighthouses. They are thinner and lighter than traditional lenses and admit more light.

when using concentrators combined with multi-layered heterostructures. The cooperation recently initiated between Kvant and Ioffe may result in major progress in this field. Thirdly, at the moment the polycrystalline silicon needed to make monocrystalline silicon for solar panels is not only expensive, but also hard to obtain. With concentrators, it is possible to use less silicon, or to use other more expensive but more efficient materials.

Another important area of research among Russian actors working to resolve the problems involved in silicon supplies is thin film solar cells, where Kvant is also active.

Production of solar cells and modules in Russia

By the late 1990s, the profile of Russia's solar power cluster was largely established. The main companies that form the cluster are listed in a table in the appendices. The overview was gathered by the Institute of Semiconductor Physics of the Siberian Branch of the Russian Academy of Sciences.[6]

All the companies sell their products almost exclusively abroad. Demand is good, in particular the European Union, where Germany constitutes the main market due to legislation and subsidies that secure the development of solar power.

Due to the heavy international demand for their solar power products, Russian producers are trying to expand their production. Those companies that are privately owned are expanding most rapidly and successfully. Most of them were founded by individuals who had worked in this or closely related spheres during the Soviet period, and were able to gain control over whole institutions or key equipment during the chaotic 1990s.

Now those companies are expanding. In 2005, Russia's output of solar elements and modules was about 12.6 MW, somewhere between 0.75 and 0.90 per cent of global production.[7] According to the Russia–EU Technological Centre, Russia by 2007 stood for approximately one per cent of world production, about 14 MW.[8]

Russia's production of solar elements and modules is concentrated in three distinct geographical locations: Krasnodar Krai (in the south of the country on

---

6 Institute of Semiconductor Physics. 'Institute of Semiconductor Physics Title Page', Siberian Branch, Russian Academy of Sciences. http://ecoclub.nsu.ru/altenergy/common/table2.htm [accessed 15 May 2008].

7 Abercade Consulting. *Rynok fotoelektricheskikh preobrazovateley i solnechnykh moduley Rossii 2005 Goda*. Moscow: Abercade Consulting, 2006.

8 Russia–EU Technological Centre, cited in Abercade Consulting 2006.

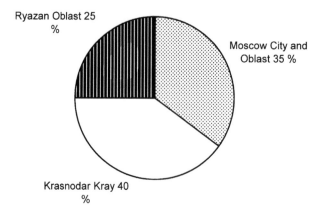

Ryazan Oblast 25 %

Moscow City and Oblast 35 %

Krasnodar Kray 40 %

Figure 4.3   Solar cells/modules produced in Russia's three producing regions, percentage of Russian output[9]

the Black Sea Coast), Ryazan Oblast (central Russia) and Moscow City and its surrounding Moscow Oblast. The biggest producer is the Krasnodar-based company Solnechny Veter. Its output of 5 MW makes up 39.7 per cent of Russia's annual production. In the city of Ryazan in west-central Russia, there are two companies engaged in the solar power sector: Soleks (founded on part of the factory Krasnoe Znamya), and the Ryazan Metal Ceramics Instrumentation Plant (RMCIP). Ryazan Oblast has a total output of 3.1 MW per year, or 24.6 per cent of Russia's total output. The remainder is located in Moscow (see Figure 4.3).

About 75 per cent of the photovoltaic elements and modules produced in Russia are made using monocrystalline silicon, with most of the remaining 25 per cent made using amorphous silicon (see Figures 4.4 and 4.5). Only Sun Energy uses amorphous silicon. Sun Energy was a joint venture of Kvant and the Russian–American company Sovlaks, but has been integrated in Kvant's other operations and is now marketed by Kvant. The advantage of polycrystalline silicon is that it makes it simpler to build solar panels, they weigh less, and less silicon is needed in their production. The disadvantage is that efficiency remains low (about 8.5 per cent as against 14 to 18 per cent for monocrystalline panels in Russia).

The relative prominence of amorphous silicon solar panels in Russia's output contrasts with the situation elsewhere, as no more than 3 per cent of global output is based on polycrystalline silicon.[10] Another peculiarity of the Russian scene is the two-sided solar modules produced by the company Solnechny Veter, with efficiencies of between 16.5 and 18 per cent.

9   Source of data used to compile figure: Abercade Consulting 2006.
10   EU–Russia Technological Centre, cited in Abercade Consulting 2006.

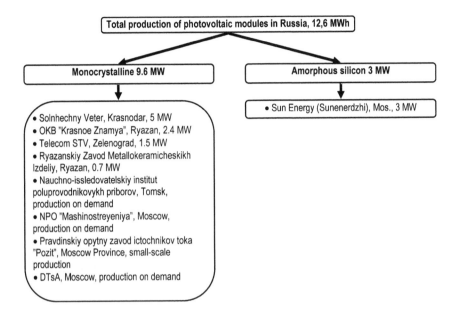

Figure 4.4 **Russian producers of amorphous and monocrystalline photovoltaic modules[11]**

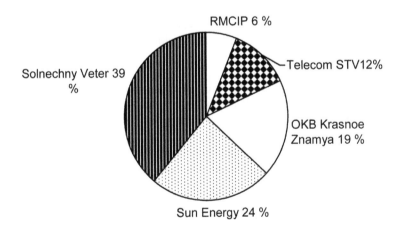

Figure 4.5 **Main Russian producers of solar modules, percentage of Russian production[12]**

---

11 Source of data used to compile table: Abercade Consulting 2006.
12 Source of data used to compile figure: Abercade Consulting 2006.

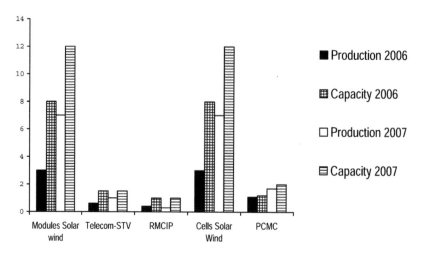

Figure 4.6   Actual production of solar cells and modules compared to production capacity, 2006 and 2007, MW[13]

Figure 4.6 shows that the actual output of the main Russian producers of solar cells and modules was significantly below their capacity. All producers have spare capacity which could be used for future production growth. The world market, towards which they are almost exclusively oriented, would be fully capable of absorbing the increased output. The problem is the deficit of polycrystalline silicon needed for producing monocrystalline silicon, which is used in over 75 per cent of solar modules made in Russia.

Until recently, all Russian producers of solar cells and modules were dependent on purchasing polycrystalline silicon on the world market, where it has become increasingly expensive as the solar power industry has begun to need far more than the scraps from the information technology industry on which it had previously relied. In some cases, production has been facilitated by barter: Customers pay with polysilicon, and get monosilicon solar cells in return, while the producer keeps a percentage of the polysilicon as its profit. Mirroring the tendency amongst Western counterparts, Russian companies are racing to develop fully vertically integrated production lines, from the production of polycrystalline silicon to the assembly of complete solar panels. Such vertically integrated operations can provide not only protection from the shortage of polysilicon in the world market but also far greater profits.

---

13   Source of data used to compile figure: Kreutzmann, Anne. 'The Smell at the End of the World: Nitol Wants to Produce Silicon in Siberia'. *Photon International*, no. 11, November (2007): 30–47.

Table 4.2   Companies planning new polysilicon factories in Russia[14]

Company	Plan
NITOL	Already started; increase to 3.7K per month from 2009
Solar Export	From 1K tonnes in 2009 to 2.5K tonnes per year later
Russian Silicon	3K tonnes per year from 2010
Renova Orgsyntes	2K-4K tonnes per year starting from 2009
Zaporozhye	3K tonnes per year starting from 2010
Kazakhstan LGK	5K tonnes per year
Kazakhstan TSK	3K tonnes per year
Synthetic Technologies	500 tonnes per year
Mining-Chemical Combine	Started manufacturing late 2007
Poldosky Plant	n.d.

The first company to succeed in establishing a new production line for the production of polycrystalline silicon in Russia is Nitol, headquartered on the outskirts of the Siberian city of Irkutsk. It has acquired a reactor for producing the silicon from the USA, and intends to sell most of its output to the Chinese company Suntech Power Holdings and the American company Evergreen Solar. Nitol is widely expected to take over as one of the biggest players in Russia, possibly also in the world. A planned initial public offering of its stock on the London Stock Exchange (LSE) was postponed due to the global credit crunch which began in late 2008, but the company was still scheduled as of mid-2009 to be registered on the LSE and continue to expand rapidly.

A further seven Russian and two Kazakh companies are also working on their own production lines for polycrystalline silicon, and expect to go online during the coming years (see Table 4.2). With most of these new factories online, the face of Russia's solar power industry will change.

The development of new production capacity for polycrystalline silicon is also a prospective area for joint ventures with Western companies. So far American, Chinese and German companies have been capturing most of the opportunities.

---

14   Sources of data used to compile table: Semiconductor Equipment and Materials International (SEMI). *Russia Market Update*, 26 February 2008. http://content.semi. org/cms/groups/public/documents/events/p042796.pdf [accessed 25 May 2008]; Wilson, cited in Business Press. 'Solnechnaya energetika rasshevelila rossiyskiy biznes'. *Novye Tekhnologii*, vol. 55, no. 428 (11 April 2008). http://businesspress.ru/newspaper/article_ mId_37_aId_446548.html [accessed 25 May 2008]; Business Press. 'Solnechnaya Energetika'. *Energeticheskoe Prostranstvo*. N.d. http://www.energospace.ru/2008/05/22/ solnechnaja-jenergetika.html [accessed 25 May 2008].

Table 4.3   Cost of solar equipment produced in Russia[15]

Company	Production, MW	Strength of solar cell, W	Strength of module, W	Efficiency of solar cell	Efficiency of module	USD/W (solar cell)	USD/W (module)
OKB Krasnoe Znamya	2.4		50–150	14–16	15	2.17–2.32	4–4.50
Telecom STV	1.5	1.43–1.50	0.75–110	15–18	14–16	–	5.00
Solnechny Veter	5.0	2.00–3.45	5.00–200	–	15–18	–	5.18–8.20
RMCIP	0.7	1.25–2.30	5.00–100	13–14	–	3.80	0.50
Sun Energy	3.0	–	7.50–150	–	8–8.5	–	10.00

The main comparative advantage of the Russian solar power industry is that its products combine relatively high efficiency with low costs, as Table 4.3 shows. Soleks (formerly OKB Kranoe Znamya) is the cheapest producer. The products of Telecom STV are also cheaper than those of its Western European competitors. This also applies to Solnechny Veter and RMCIP, depending on the product in question.

---

15   Source of data used to compile table: Abercade Consulting 2006.

Chapter 5
EU–Russian Science and Energy Cooperation[1]

The examination of energy relations between the European Union and Russia is of special interest to this study. The EU is, in spite of wrangling over various issues, Russia's main international trade partner and interlocutor; moreover, several EU member countries share long borders with Russia and are thus directly or indirectly affected by Moscow's regional policies, including those related to energy development and exports. With Russia's resurgent economic growth during the past decade and the growing importance of its energy reserves, building a 'solid strategic partnership' with Russia has become a key policy goal of the EU.[2] Russia is a major geopolitical player both regionally and globally, while its geographical proximity to the EU, its role as an energy supplier to the EU and as a market for EU goods and investments make good relations essential to both sides. The EU is a key market for Russian oil and gas, an important trade partner and a valuable source of investment. The level of economic interdependence may have heightened sensitivity and vulnerability to each other's economic policies, but these linkages have not yet translated into significantly closer political ties.

This chapter surveys existing EU–Russian cooperation in the field of renewable energy and draws some general conclusions on what factors are conducive to successful joint projects, and what should be avoided by other actors seeking to develop similar cooperation in the future. Despite, or perhaps also because of, the level of economic interdependence between the EU and Russia there have been tensions in the relationship. In that context, renewable energy may prove to be an area of 'low politics' in which cooperation can be developed fruitfully, providing a basis for a better mutual understanding between the two sides.

EU–Russian relations

A common EU policy on Russia was formulated in June 1999 but lapsed in 2004 when it became apparent that the EU's member countries had divergent ideas about

---

1   This chapter was co-authored by Grant Dansie.

2   European Commission Delegation to Russia. 'Overview of Relations', 2007, EC Delegation. http://www.delrus.ec.europa.eu/en/p_210.htm [accessed 3 June 2008].

how best to engage Moscow.[3] As is often the case with the EU in foreign policy, there are considerable difficulties in finding common ground due to the diversity of member-state interests. With regard to energy, there is a heavy dependence on Russian energy imports among eastern member states like Poland, which imports 62 per cent of its gas from Russia, and the Baltic states, which depend on Russia for 100 per cent of their gas supply. Germany imports 43 per cent of its gas from Russia, but otherwise in Western Europe there is much less dependence on Russian energy supplies.[4] This, along with the fact that much of the gas transported to Europe is transited through Ukraine, shapes European policy responses to events like the repeated gas disputes between Ukraine and Russia in 2006 and 2009.

In 2006, a dispute over gas pricing between Russia and Ukraine led to a temporary suspension of gas supplies via Ukraine,[5] adversely affecting other Eastern European EU members. A similar drama unfolded in early 2009 when a dispute over prices led to another two-week cut-off of gas to Ukraine, with other European states being affected.

The recurrent disputes between Russia and Ukraine over gas pricing have exacerbated the often frosty political relations between the two former Soviet states in the wake of the Orange Revolution of 2004, which swept in the pro-Western government of Viktor Yushchenko and drew Moscow's ire. By June 2009, the European Commission was offering to mediate between the two sides as yet another disagreement over gas trade prices threatened to trigger further punitive measures from Russia.[6] Russian gas has also been a factor in the conflict between Georgia and Russia, as Russia cut gas supplies to Georgia in 2006. The disputed territory of South Ossetia, over which the two sides fought briefly in August 2008, has continued to receive Russian gas.[7] European states responded to these events with great alarm and concern, fearing that Russia was not only seeking a more assertive policy in its former sphere of influence, but was also backing it up militarily.

The most serious impact of the Russian disputes with Ukraine on EU–Russian relations is probably the major shift in European perceptions of Russia as an

---

3   Massari, Maurizio. 'Russia and the EU Ten Years On: A Relationship in Search of Definition'. *The International Spectator*, vol. 42, no. 1, (March 2007): 1–15. http://www. iai.it/pdf/articles/massari.pdf [accessed 21 June 2009]. p. 4.

4   *Russian Analytical Digest*, 'Russia's Energy Policy', no. 18 (3 April 2007): 1–17, p. 13.

5   Stern, Jonathan. 'The Russian–Ukrainian Gas Dispute', Oxford Institute for Energy Studies, 16 January 2006. http://www.oxfordenergy.org/pdfs/comment_0106.pdf [accessed 22 June 2009].

6   Agence France-Presse. 'EU to Send Mission to Moscow and Kiev over Gas Dispute', 4 June 2009. http://news.id.msn.com/topstories/article.aspx?cp-documentid=33 58147 [accessed 21 June 2009].

7   Gearóid Ó Tuathail, 'Russia's Kosovo: A Critical Geopolitics of the August 2008 War over South Ossetia'. *Eurasian Geography and Economics*, vol. 49, no. 6 (2008): 670–705.

energy supplier. Although there were few reasons to believe that Russia would treat any of the customers within the EU in the same way as it has treated countries in its perceived sphere of interest, the question of Russia's reliability as an energy supplier arose, and became linked with the question of whether Europe should partner with an increasingly authoritarian Russia.[8] This situation also strengthened the EU's resolve to diversify its energy supplies, and during the same time period the international focus on climate change peaked. In January 2008, the EU committed to the '20/20/20' goals. This is a package of proposals designed to reduce EU emissions to at least 20 per cent below 1990 levels by 2020, while increasing the share of renewable energy to 20 per cent during the same period.[9] However, these initiatives have not diminished Moscow's importance in Europe's short- and medium-term energy policies. In fact Europe may become more reliant on Russian gas, as natural gas emits less CO_2 than do coal or oil. Even with the current negative spiral in relations due to the tensions over Russia's sometimes stormy relations with other post-Soviet states, it is difficult to see how Russia and the EU can isolate themselves from each other when it comes to energy trade.

The legal basis for EU–Russian relations has been the Partnership and Cooperation Agreement (PCA), which entered into force in 1997 and was designed to last ten years. The PCA has been based on mutuality, and aims to strengthen political, economic, commercial and cultural ties between the EU and Russia. It provides an institutional framework for cooperation by facilitating consultation and dialogue on several levels, including heads of state summits, ministerial meetings, interparliamentary dialogue and high-level bureaucratic interaction.[10] In May 2003, the EU and Russia sought to strengthen their cooperation further by establishing four 'common spaces' within the framework of the PCA:

i. The Common Economic Space
ii. The Common Space on Freedom, Security and Justice
iii. The Common Space on External Security
iv. The Common Space on Research, Education and Culture

The PCA expired at the end of 2007, and attempts by both sides to create a successor agreement were hampered the following year as a result of the conflict between Russia and Georgia over South Ossetia. Efforts to construct some sort of successor agreement have also been adversely affected by lack of agreement among EU members over a common foreign policy. At the same time, the Russian

---

8   Perović, Jeronim and Robert Orttung. 'Russia's Energy Policy: Should Europe Worry?'. *Russian Analytical Digest*, no. 18 (2007): 2–7.

9   EC. 'The Climate Action and Renewable Energy Package, Europe's Climate Change Opportunity', 8 January 2009. http://ec.europa.eu/environment/climat/climate_action.htm [accessed 21 June 2009].

10   EC 2009.

side is wary of what it perceives to be EU attempts at integrating states within Russia's own sphere of influence into EU. For example, Russian President Dmitri Medvedev expressed his frustration with the state of EU affairs at a May 2009 Summit in the far-eastern Russian city of Khabarovsk. He criticized what he saw as EU attempts to construct an Eastern Partnership Plan which would draw the Union economically closer to other former Soviet states – a move Medvedev saw as a veiled attempt to constrain Moscow's policies in former Soviet republics.[11] It is in this atmosphere of political uncertainty that both sides nonetheless have had to develop a comprehensive energy dialogue – also covering renewable energy.

EU–Russia Energy Dialogue

A joint declaration on the EU–Russia Energy Dialogue was signed during the sixth EU–Russia summit in Paris in October 2000. According to the declaration, the aim of the dialogue was to 'enable progress to be made in the definition of an EU–Russia energy partnership and arrangements for it'.[12] It was hoped that this would provide an opportunity to raise energy-sector questions of common interest on a regular basis. Discussions were to take place on a wide range of subjects, but the main focus of the dialogue has been on challenges related to the strong interdependence of the EU and Russia in the energy sphere. This mutual dependence is primarily linked to Russia's natural abundance in hydrocarbons, which account for the lion's share of its exports to the EU. Some 53 per cent of the oil exported from Russia goes to the EU, as well as 36 per cent of gas exports. This accounts for 21 per cent of the EU's net oil imports and 41 per cent of its gas imports.[13]

The strong focus on petroleum resources notwithstanding, renewable energy and energy efficiency are also covered by the dialogue. Among the first projects to be started were pilot projects on rational energy use and savings in the Arkhangelsk and Astrakhan regions,[14] later expanded to include the Russian enclave of Kaliningrad. Also, the EU–Russia summit in 2001 recognized the need to study the possibilities of common implementation of energy-saving and renewable projects.

---

11   Agence France-Presse. 'EU–Russia Summit Fails to Mend Rifts', 27 May 2009. http://news.yahoo.com/s/afp/20090522/wl_afp/russiaeusummit_20090522183530 [accessed 21 June 2009].

12   Chirac, Jacques, Javier Solana, Romano Prodi and Vladimir Putin, 'Joint Declaration', 10 October 2000. Brussels: EC. http://europa.eu/rapid/pressReleasesAction. do?reference=IP/00/1239&format=HTML&aged=0&language=EN&guiLanguage=en [accessed 21 June 2009].

13   Piper, Jeff. 'Toward an EU–Russia Energy Partnership'. Presentation at the conference *Energy Security: The Role of Russian Gas Companies*, Paris, France, 25 November 2003.

14   EU. 'Issues Being Examined Under the Energy Dialogue'. 2007. http://ec.europa. eu/energy/russia/overview/issues_en.htm [accessed 7 June 2008].

The aim was to make a catalogue of projects of this type in Russia that could be financed under the joint implementation mechanism of the Kyoto Protocol, an initiative which seemingly has yet to materialize. Energy efficiency, in contrast to renewable energy, became a priority area for the Energy Dialogue and was the subject of one of three thematic groups under the Energy Dialogue (the two others being Energy Scenarios and Market Developments). The EU side in the Energy Dialogue warmly welcomed Russia's ratification of the Kyoto Protocol in 2003. It noted that this opened opportunities to 'engage in joint implementation (JI) projects and in emissions trading', as well as being an encouragement to make the 'additional efforts necessary to improve energy efficiency and to develop non-polluting energies, notably renewable energy sources'.[15] Over time, however, the Kyoto Protocol has received less attention in the Energy Dialogue.

The dual gas pricing system in Russia, with differentiated prices for domestic consumers and for exports, has been a major stumbling block in the Energy Dialogue.[16] The EU views the system as a *de facto* trade barrier which gives Russian energy-intensive industries unfair advantages over their EU counterparts, and this has contributed to European reservations about allowing Russia into the World Trade Organization despite many years of negotiations. Moscow has argued that the EU's advocacy against the dual pricing system is an attempt to deprive Russia of one of its natural competitive advantages, namely its abundance of natural gas. It has also noted that, since there is no agreed global price standard for natural gas, this issue should be de-coupled from WTO negotiations – a stance which Brussels has rejected.[17] Since subsidies for natural gas in Russia are seen as the main impediment to the development of renewable energy (as detailed in Chapter 2), the focus on energy subsidies in the Energy Dialogue indirectly concerns renewable energy, although renewable energy *per se* is rarely discussed.

One exception is the EU–Russia Technology Centre, also known as OPET Russia, which was established in November 2002 as part of the Energy Dialogue, and under the auspices of the Work Programme for Energy of the 5th Framework Programme of the EU. This was one component in the broader network Organizations for the Promotion of Energy Technologies (OPET), which aimed to

---

15 Khristenko, Victor and François Lamoureux. 'EU–Russia Energy Dialogue: Fifth Progress Report'. Brussels: EC, 2004. http://www.ec.europa.eu/energy/russia/joint_progress/doc/progress5_en.pdf [accessed 8 June 2008], p. 2.

16 Skurbaty, Tim. *Understanding the EU–Russia Energy Relations: Conflictual Issues of the ED and the ECT.* MA thesis, University of Lund, Department of Political Science, 2007. http://theses.lub.lu.se/archive/2007/05/16/1179319221-4953-154/MEA_thesis.pdf [accessed 7 June 2008]. p. 35.

17 Ripinsky, Sergey. 'The System of Gas Dual Pricing in Russia: Compatibility with WTO Rules'. *World Trade Review*, vol. 3, no. 3 (November 2004): 463–81. p. 464.

further the deployment of innovative technologies and increase the pace of market uptake in respect of research that supports European Energy Policy priorities.[18]

Against this backdrop, the EU–Russia Technology Centre was to provide a forum for the exchange of ideas and information, and facilitate a structured framework for cooperation. The Centre was to provide training for specific target groups on energy technologies, technical assistance for the introduction of advanced energy technologies and undertake information dissemination and communication activities. Activities were divided by sector – hydrocarbons, coal and electricity, renewables, energy saving and efficiency.[19] With regard to renewable energy, the Centre's work resulted in a report on the potential of renewables in Russia and the available technologies.[20] However, there seems to have been little follow-up of this report. The joint progress reports on the Energy Dialogue have not mentioned the Kyoto Protocol or renewables since the publication of the Fifth Progress Report in 2004. Although its website continues to exist on the Internet, the EU–Russia Technology Centre was shut down as of 2005. A competitive call was made for organizations that wanted to continue its work. A new Russian organization won this call, but thus far no new centre has come into existence. However, various other mechanisms have facilitated energy cooperation between Europe and Russia, with varying degrees of success. We now turn to these.

Cooperation under FP7

The EU's Seventh Framework Programme (FP7) was launched in 2007 and is expected to continue operating until 2013. Its purpose is to provide and coordinate grants for various European projects in science and technology related to EU interests.[21] The FP7, with a budget of about EUR 50 million, includes over 200 joint Russian–European projects. However, it should be noted that these projects are largely dependent on EU funding, and most are not related to energy. Energy is nonetheless one of the central facets of the current FP7 Cooperation Work Programme, where Russia plays an important role. The overall objective in this area is:

> adapting the current energy system into a more sustainable one, less dependent
> on imported fuels and based on a diverse mix of energy sources, in particular

---

18   EU. 'Introducing the OPET Network', N.d. http://cordis.europa.eu/opet [accessed 9 June 2008].

19   Khristenko and Lamoureux 2004.

20   EU–Russia Technology Centre. *Renewable Energy Sources Potential in the Russian Federation and Available Technologies*. Moscow: EU–Russia Energy Technology Centre, 2004.

21   EC. 'FP7 in Brief: How to Get Involved in the EU 7th Framework Programme for Research'. Brussels: EC, 2007c. http://www.ec.europa.eu/research/fp7/pdf/fp7-inbrief_en.pdf [accessed 21 June 2009].

renewables, energy carriers and non-polluting sources; enhancing energy efficiency, including by rationalizing use and storage of energy; addressing the pressing challenges of security of supply and climate change, whilst increasing the competitiveness of Europe's industries.[22]

Following on from this, the research, development and demonstration activities carried out under this Work Programme are expected to:

- improve energy efficiency throughout the energy system, taking into account the global environmental performance;
- accelerate the penetration of renewable energy sources;
- de-carbonize power generation and, in the longer term, substantially de-carbonize transport;
- reduce greenhouse gas emissions;
- diversify Europe's energy mix;
- enhance the competitiveness of European industry, including through a better involvement of small and medium enterprises (SMEs).[23]

In terms of EU–Russian collaboration, the EU concluded the 'Energy EU Russia Call' in February 2008. This call was centred on two key activities: enhancing strategic international cooperation with Russia in the field of power generation from biomass; and Pan-European Energy Networks, which will involve the development of innovative operational and monitoring tools for large power systems. These are both collaborative projects funded jointly by the European Commission and Russia's Federal Agency for Science and Innovation, with each partner dedicating a budget of EUR 4 million to the two projects. Additionally, the European Community may contribute to the Russian participants up to 5 per cent of their total eligible costs. This contribution will only cover costs not funded by the Federal Agency for Science and Innovation.[24] As far as we have been able to ascertain, the selection of biomass as one of the focal topics is a result of lobbying by Finnish and Swedish actors in the forestry sector. It is thus an example of the centrality of the Nordic countries to EU–Russian cooperation on renewable energy. The co-funding of these collaborative projects is also indicative of Russia's increasing willingness and ability to contribute financially to scientific cooperation with international actors. Other Russian–EU collaboration includes work on research excellence and major infrastructure in Russia, and the potential for science and technology cooperation with EU partners in the area of the environment.

---

22   EU Work Programme. 'Cooperation Theme 5 Energy', 2008. ftp://ftp.cordis. europa.eu/pub/fp7/docs/wp/cooperation/energy/e_wp_200902_en.pdf [accessed 21 June 2009]. p. 4.

23   EU Work Programme 2008, p. 5.

24   EU Work Programme 2008, p. 9.

EU–Russian cooperation under the Kyoto Protocol

With the increased focus on climate change and the ratification of the 1997 Kyoto Protocol have come several policy changes in EU countries regarding how they approach renewable energy development in Russia, which acceded to the protocol in 2004. Under the terms of the Kyoto Protocol, the EU committed to making an 8 per cent reduction in greenhouse gas emissions in the period 2008–2012 compared to the 1990 level. To reach these goals the EU introduced a system of tradable emission permits. However, the Flexible Mechanisms developed under the protocol offer alternative ways in which to reduce emissions, via investments in emission-reduction projects in industrializing or developing countries. These mechanisms are Joint Implementation (JI) and the Clean Development Mechanism (CDM). The idea behind the JI mechanism is to enable industrialized countries with a greenhouse gas reduction commitment to fund emission-reducing projects in other industrialized countries, as an alternative to emission reductions in their own countries. The JI mechanism can be applied when both countries in question have ratified the Kyoto Protocol. The majority of JI projects thus far have been undertaken in countries in Eastern Europe. The CDM on the other hand is used when one country has signed the Kyoto Protocol, and the other has not. It is therefore more oriented towards developing countries and is less relevant for Russia. These mechanisms have allowed net global greenhouse gas emissions to be reduced at a lower global cost by financing emissions reduction projects in countries where costs are lower.[25]

Russia offers a prime opportunity for the implementation of these Flexible Mechanism projects due to its position as one of the most energy-intensive economies in the world, and one of the most inefficient and wasteful. Indeed it has been estimated that energy conservation potential in Russia represents some 40–45 per cent of its current energy consumption.[26] If one adds to this EU expertise in utilizing renewable energy, there are various benefits to be achieved. EU countries will be better able to fulfil their commitments to reducing emissions under the Kyoto Protocol, while at the same time establishing a presence in Russia's potential renewable energy market. This presence may have significant benefits for the companies involved as they can effectively demonstrate their expertise, while at the same time establishing a name for themselves in a rapidly evolving market.

---

25   United Nations Framework Convention on Climate Change website. http://unfccc. int/kyoto_protocol/mechanisms/items/1673.php [accessed 25 August 2008]. The debate on whether or not these mechanisms actually do contribute to emission cuts is outside the scope of this report, but for counter-arguments see Vetlesen, Arne Johan. 'Nullsumlogikk og andre narrespill–norsk oljepolitikk i klimakrisens tjeneste'. *Samtiden*, 2, 2008: 54–64.

26   Makarov, Alexey. *New Energy Consumption and Supply Trends (Worldwide and Russia)*. Moscow: Energy Research Institute, Russian Academy of Sciences, 2004, p. 3.

Many EU actors are aware of this, and several EU agencies are now channelling their bilateral renewable energy cooperation through these Flexible Mechanisms, and specifically towards Russia. Of the 163 JI projects currently underway in the world, 109 are located in Russia, the majority of them initiated by actors in the EU countries.[27] The next big challenge for EU/Russian cooperation will come in December 2009 during the Copenhagen Summit, which is expected to produce a successor treaty to the Kyoto Protocol. Large emerging markets like Russia's will be under great pressure during these negotiations.

INTAS

One of the most important initiatives in EU–Russian scientific cooperation was the establishment of INTAS, the International Association for the Promotion of Cooperation with Scientists from the Newly Independent States of the Former Soviet Union. Although not focusing exclusively on energy, INTAS was the most comprehensive cooperation programme with Russia. As it ended in 2006 and has been subjected to a thorough evaluation, it can offer several important lessons for others wishing to promote research cooperation with Russia. INTAS emerged in 1993 as a flexible response to the dire situation faced by researchers in the newly independent states (NIS) after the economic collapse of the Soviet Union.[28] Its primary aim was to fund research and foster cooperation between the international scientific community and the former Soviet republics, in particular scientists who had been involved in military applications and whose competence might be misused. INTAS was one of several bi- and multilateral funding schemes aimed at promoting science and technology in the former Soviet republics. Although not the largest such initiative, it is probably the best known. INTAS ended along with the 6th EU Framework Programme (FP6) in 2006.[29]

Between 1993 and 2003, INTAS funded 2,726 projects, bringing together 50,000 participants in 15,000 research teams. During this period INTAS funded projects worth EUR 220 million. The sources of this funding were diverse, with EUR 194.5 million (88 per cent) coming from the EC, EUR 15 million (7 per cent) from various member countries and EUR 10.5 million Euros (5 per cent) from

---

27   UNEP/RISØ. 'JI Projects'. http://cdmpipeline.org/ji-projects.htm [accessed 14 August 2008].

28   INTAS. *Report by External Evaluators on the Programme of the International Association for the Promotion of Co-operation with Scientists from the New Independent States of the Former Soviet Union (INTAS) in the Period 1993–2003 to the INTAS General Assembly*, Brussels, INTAS, 1 October 2004.

29   European Commission Delegation to Russia. *The EU and Russia: Exploring Beyond Borders*. Moscow, EC Delegation, 2006. http://www.delrus.ec.europa.eu/en/images/mText_pict/2/science%20eng.pdf [accessed 20 June 2009], p. 13.

various former Soviet republics. Russia's share of INTAS was significant, with over 70 per cent of funds being allocated to Russian research groups.[30] In order to disseminate information about its funding mechanisms and activities, INTAS operated 15 information desks throughout the former Soviet republics.[31] These were criticized in the 2004 evaluation report of INTAS for their low visibility, which was most probably due to lack of resources.[32] In 2002 INTAS began supporting the participation of NIS researchers in the EU FP, which resulted in the establishment in 2003 of the FP6 NIS Information Network (ININ). In the 2004 evaluation report, INTAS was again criticized for the poor communication and cooperation between these agencies, which were later merged.[33]

The evaluation panel concluded that INTAS had been largely successful in fulfilling its objective of promoting scientific cooperation between NIS scientists and the international scientific community. For many NIS scientists, INTAS provided their first experience of international cooperation, becoming their principal means of remaining in academia.[34] However, the panel also concluded that although INTAS had effectively preserved and maintained the existing research infrastructure, it had been less successful in triggering major and much-needed reforms of the NIS national innovation systems. Overall, the achievements of INTAS during its 1993–2003 lifespan were impressive, with 18,000 publications in international refereed journals, 22,000 presentations at international conferences and workshops, and 500 patents.[35] The EC decided to discontinue INTAS with the conclusion of FP6, but many of its activities have now been integrated into FP7.[36] In addition, many scientists who participated in joint projects have later worked together outside the auspices of INTAS, highlighting the long-term benefits of cooperation.[37]

The INTAS experience demonstrated the usefulness of several incentives that can be utilized to attract the most serious Russian researchers and research institutions to joint cooperation projects. Firstly, tax-free grants and duty-free imports of equipment can be attractive. A further significant incentive lies in the

---

30   INTAS 2004, p. 20.
31   Runge, Tatiana. *ININ Results, Achievements and Recommendations.* Brussels, ININ, 28 November 2006. http://www.intas.be/documents/ininworkshops/6inin_ workshop_presentations/ININ_results_achievements_&_recommendations.pdf [accessed 20 May 2008], p. 3.
32   INTAS 2004, p. 8.
33   INTAS 2004, p. 8.
34   INTAS 2004, p. 9.
35   INTAS 2004, p. 9.
36   British Council (2007) *British Council Online Bulletin*, September 2007. http://64.233.183.104/search?q=cache:YEYdTBDJDO0J:www.ukro.ac.uk/insight/ei0709. doc+why+was+INTAS+discontinued&hl=en&ct=clnk&cd=7&client=firefox-a [accessed 30 May 2008].
37   INTAS 2004, p. 9.

opportunity to work with highly advanced equipment not available in Russia. The INTAS evaluation panel found that most proposals were initiated by an NIS partner, which pointed up the importance of making any international initiatives highly visible to the Russian scientific community. If this is to be achieved through helpdesks or similar facilities, the low impact of INTAS helpdesks underlines the importance of making them more visible. Streamlining funding opportunities under a single or few entities could be an effective measure, reducing bureaucracy and at the same time making funding sources more visible. The popularity of INTAS also shows the importance of creating a good and lasting brand name that both private and public actors can become familiar with and can remember.

Flexibility can be an essential element in effective cooperation. The 2004 evaluation highlights one of the positive aspects of INTAS as being the ability to transfer money from one budget post to another within projects – something that is often difficult with other European research funding.[38] An additional success was the funding of individuals in some cases, without necessarily having to involve national governments or institutions. On the other hand, the evaluation also noted that the quota system which placed limitations on the number of researchers from each state had been a significant hindrance, reducing commitment and initiative.[39] The evaluation panel also highlighted some of the problems that INTAS encountered in Russian research and innovation policy:[40]

- uneven policy formulation processes and the overall marginal status of science and research in governmental policy priorities (a greater problem in the 1990s than it is currently);
- weak policy delivery systems in terms of institutional structures and capabilities;
- lack of sufficient public budget allocations for research and technical development;
- timid reform of the Academy of Sciences and the corresponding struggle in the competition for funding;
- generational change (the retirement of the 'Sputnik' generation), related problems and the problem of internal and external brain drain in Russia.

Although the situation in today's Russia is somewhat different from the early years of the INTAS programme, many of the issues highlighted may still have relevance. Russia has three major assets: a vast wealth of natural resources, a highly educated population (particularly in the natural sciences and technology), and proximity to Western Europe.[41] Effective international cooperation with Russia presents a

---

38 INTAS 2004, p. 25.
39 INTAS 2004, p. 9.
40 INTAS 2004, p. 33.
41 INTAS 2004, p. 7.

range of opportunities for utilizing these assets, opportunities that can benefit the scientific knowledge base as well as the economic standing of both parties.

IIASA

The International Institute for Applied Systems Analysis (IIASA) is an international research organization located in Laxenburg, Austria. Founded in 1972, its research focuses on various aspects of environmental, economic, technological and social issues in the context of global change. IIASA's research is organized around fields of policy importance rather than academic disciplines; the institute's investigators perform interdisciplinary research that combines methods and models from the natural and social sciences in addressing areas of concern to all societies.[42] Historically, IIASA has constituted a bridge between Western and Soviet/Russian research. During the Soviet period, *perestroika* in the 1980s and the post-Soviet period, the institute has included many Russian researchers and has worked closely together with the Russian Academy of Sciences and other Russian research organizations. The overall objective of IIASA's Energy Programme, initiated in March 2006, is to better understand the nature of alternative future energy transitions, their implications for human well-being and the environment, and how they might be shaped and directed by current and future decision-makers. Its three major areas of activity consist of coordinating the Global Energy Assessment, developing new methods and modelling techniques for exploring alternative energy pathways, and longer-term research on energy investment requirements.[43]

IIASA scientists are actively involved in a number of international energy-related studies and networks. These range from the IPCC Fourth Assessment Report (with four lead authors and one coordinating lead author on various aspects of energy-related climate change mitigation), the United Nations Sigma Xi Scientific Advisory Group on Energy and Climate to the United Nations Commission on Sustainable Development (UN Sigma Xi 2007), the InterAcademy Study on Energy and Climate (InterAcademy 2007), the International Council for Science (ICSU) Study on Energy and Sustainable Societies (ICSU 2004) as well as the new ICSU initiative on energy-related research.

Joint projects in Russia: The view from Murmansk

As a part of this project, a database of international renewable energy projects in Russia has been compiled, currently covering 83 projects. It is incomplete, and we

---

42   International Institute for Applied Systems Analysis (IIASA) homepage. http://www.iiasa.ac.at/docs/Research/ [accessed 8 August 2008].

43   IIASA homepage 2008.

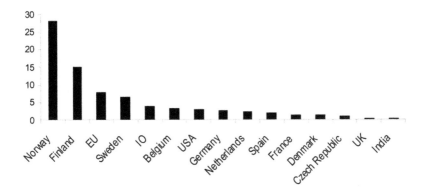

Figure 5.1   **Number of renewable energy/energy efficiency projects in Russia by foreign country**[44]

intend to continue developing it in the future. According to our data, illustrated in Figure 5.1, Norway is by far the biggest Western actor in terms of cooperation with Russia on renewable energy and energy efficiency. Finland comes in as a good number two. It is worth noting that the Nordic countries jointly stand for well over half of the activity.

It is possible, though far from certain, that these data have firstly a Norwegian, and secondly a Nordic bias. Although the data were gathered primarily by a Russian living in Russia, the centre of the data collection was in Murmansk, where both Norway and Finland have a strong presence. Perhaps these initial findings can serve as a challenge to representatives of other countries to release more information about their projects. In future versions of this graph it is expected that Japanese and Chinese projects can be added, and the German and perhaps US columns may grow considerably. Nonetheless, the Nordic countries are clearly major actors in Russia's renewable energy sector.

Conclusions

This chapter has examined the seeming divergence between high-level governmental dialogue between the European Union and the Russian Federation in areas related to energy cooperation, and various initiatives on the sub-state and non-governmental levels. While the former has been adversely affected by geopolitics and diplomatic differences, the latter set of initiatives has produced more tangible results as well as the basis for further developing ties. Despite differences between

---

44   'IO' stands for World Bank or EBRD. Where several countries are involved in the same projects, they have been ascribed equal shares of the project (e.g. 0.3).

Moscow and the EU over Russia's energy policy and continued concerns in the Russian government over Europe's intentions towards Russia, the two sides are still bound together in energy dependency and must therefore continue to find ways to reconcile their political differences. Individual initiatives from specific EU states and other actors have mitigated these problems, but the current lack of a common Russia policy and a common energy policy among EU member countries will also be an obstacle to further effective engagement with Moscow.

The drop in global oil prices since their peak at USD 147 per barrel in July 2008 has also shifted the power dynamics between the two sides. Russia has been hit especially hard by the double blow of volatile energy prices and the global recession, and these effects have spilled over into its gas production and exports.[45] With Moscow in less of a position to take a confrontational approach towards Brussels, both sides might be better situated to build consensus. However, the regional complications surrounding Georgia and Ukraine, as well as the likely delicate political manoeuvring leading up to the Copenhagen Summit in December 2009, will continue to provide challenges, underscoring the need for linkages on a variety of levels designed to encourage deeper cooperation and engagement in joint energy development. As pointed out in the introductory chapter to this book, renewable energy development may serve as a fruitful 'low politics' area in which it could be easier to build mutual trust than in the often foreign policy-entangled hydrocarbons trade. In light of previous difficulties in EU–Russian relations, this would seem to be something that European policy-makers should consider in the development of future cooperative channels with Russia.

---

45   Medetsky, Anatoly. 'Gazprom to Delay Field Due to Low Demand'. *Moscow Times*, 16 June 2009. http://www.themoscowtimes.com/article/1009/42/378835.html [accessed 21 June 2009].

Chapter 6

Nordic–Russian Cooperation on Renewable Energy[1]

This chapter surveys key bi- and multilateral actors involved in Nordic–Russian cooperation in the field of renewable energy. This information is aimed at helping other international actors gain access to the lessons learned. Cooperation between the Nordic countries and Russia is of interest to an international audience for several reasons: the Nordic countries are world leaders in renewable energy technology; the Nordic region, unlike the USA, Japan or most EU countries, shares a long land border with Russia; and Nordic companies have invested heavily in the Russian economy, more so on average than the companies of other OECD countries. For all these reasons, the richest experience of renewable-energy cooperation with Russia is to be found in the Nordic countries, and can provide important lessons for others.

The chapter outlines the forms of cooperation taking place in the areas of research and innovation on renewable energy, environmental studies, hydrogen, climate change and energy efficiency, albeit with a clear focus on renewable energy. We will look at the structure of funding, the interrelationships between the various organizations, and the types of funding mechanisms available. This will be done on a country-by-country basis, followed by an overview of multilateral cooperation in the above-mentioned areas.

In preparing this chapter, one point that quickly became clear to us was the interwoven nature of various Nordic cooperative initiatives towards Russia. The individual Nordic countries utilize multilateral organizations as their main instruments for channelling funds, while the multilateral organizations have deep levels of cooperation and interaction with each other. Thus, this chapter focuses mostly on multilateral actors, as the current tendency is for countries to shift gradually from bilateral to multilateral cooperation. This will be particularly evident in the case of Sweden.

A large number of pan-Nordic and pan-Baltic institutions have been created to foster cooperation. It also appears that there is significant overlap between the goals and aims of the various multilateral organizations, as will become evident later in the chapter. The chapter ends with a mapping of the comparative advantages of Russia and each of the Nordic countries, revealing complementarities that could

---

1 This chapter was co-authored by Grant Dansie.

be utilized to maximize the output from cooperation. This exercise could be useful for actors from other countries seeking to engage in renewable energy partnerships with Russia, in order not to let their own geographical and technological biases overly shape the cooperation in a way that might result in the partners missing out on comparative advantages. But first we will examine the nature of bilateral cooperation, starting with Denmark.

Denmark

The Danish Energy Authority

The Danish Energy Authority is part of the Ministry of Climate and Energy. It seeks to promote Denmark's international position in the area of energy, and to strengthen the business and export opportunities for Danish energy technology and know-how. The agency promotes Danish energy policy interests through its international, multilateral and bilateral cooperation on energy and environment policy. These activities take place in a range of fora, including the EU, the European Energy Charter, the Organization for Economic Cooperation and Development (OECD), the International Energy Agency (IEA), the UN and the Nordic Council of Ministers, and with various bilateral cooperation partners.[2]

Most bilateral cooperation takes place through DanishCarbon, the governmental programme for supporting international projects that reduce emissions of greenhouse gases. DanishCarbon is managed by the Danish Energy Agency under the Ministry of Climate and Energy. DanishCarbon develops projects under the Kyoto Protocol's flexible mechanisms of Joint Implementation (JI) and Clean Development Mechanism (CDM), and purchases carbon credits on behalf of the Danish state.[3] Indeed, the shift to working/developing projects under the auspices of the Kyoto Protocol is a notable tendency, not only in Denmark but in many other Kyoto Protocol signatories in Western Europe.

DanishCarbon's funding mechanism incorporates both grants and credits, but is currently in a transition stage reflecting organizational and policy changes. Grants are steadily being phased out, whilst there is more and more focus on credits. In terms of identifying projects, the process has been both 'top–down' and 'bottom–up'. DanishCarbon has identified a range of relevant projects, but in 2007 it also published an open call.[4]

---

2   Danish Energy Authority. http://www.ens.dk/da-DK/Sider/forside.aspx [accessed 21 June 2009].

3   Interview with Bo Riisgaard Pedersen, Chief Programme Coordinator, Climate and Energy Economics, Danish Energy Agency, 30 July 2008.

4   Interview with Bo Riisgaard Pedersen 2008.

The Danish Agency for Science Technology and Innovation

This agency functions under the auspices of the Ministry of Science, Technology and Innovation. It has a number of separate research councils dealing with various aspects of innovation and energy policy. Of particular relevance are the Programme Commission on Sustainable Energy and Environment and the Programme Commission on Strategic Growth Technologies.

RISØ DTU, National Laboratory for Sustainable Energy

RISØ operates under the aegis of the Technical University of Denmark and is heavily involved in research into renewable and sustainable energy. It has significant contact with Russian researchers, hosts several Russian guest researchers, and has organized international conferences involving Russian actors.

RISØ does not have any funding to initiate large-scale projects, so its focus is on facilitating research and bringing together researchers, institutions and businesses. It has been active in some small projects in Russia: for instance it developed a Wind Atlas of Russia.[5]

Russian–Danish Energy Efficiency Institute

In 1994 the Danish government established the Russian–Danish Energy Efficiency Institute near Moscow to promote the transfer of energy efficiency know-how and technology between Danish and Russian companies. The institute supports meetings, exhibitions, contacts with Russian authorities, and assists with the preparation and implementation of demonstration projects.

Norway

SIU

SIU is the Norwegian acronym for the Norwegian Centre for International Cooperation in Higher Education. SIU has several initiatives aimed at fostering research cooperation with Russia, notably the Fellowship Programme for Studies in the High North and the Norwegian Cooperation Programme with Russia. In 2007 SIU announced that it would allocate NOK 42 million to various projects under the Norwegian Cooperation Programme with Russia. One of these was 'Joint training of specialists and students in the field of electro-catalysis for hydrogen energy', which was awarded NOK 1.3 million between 2008 and 2010.

---

5   See Starkov, Aleksandr, Lars Landberg, Pavel Bezroukikh and Mikhail Borisenko. *Russian Wind Atlas*. Roskilde: Risø National Laboratory, 2000.

Following this, SIU and the Norwegian Ministry of Foreign Affairs signed an agreement for the period 2007–2010 regarding a fellowship programme for studies at institutions of higher education in northern Norway. The Fellowship Programme for Studies in the High North forms part of the Norwegian government's High North Strategy.

The Norwegian Research Council (NRC)

The Norwegian Research Council's cooperation projects in Russia take place under the same mechanisms as SIU. For instance, the NRC also works under the auspices of the Norwegian Cooperation Programme with Russia – with both institutions receiving funding for projects. However, in the current project pipeline there is little emphasis on renewable energy in the Research Council's cooperation programme with Russia.

The Nansen Environmental and Remote Sensing Centre

The Nansen Environmental and Remote Sensing Centre (NERSC) is an independent non-profit research institute affiliated with the University of Bergen, Norway. The Nansen Centre conducts basic and applied environmental research funded by national and international governmental agencies, research councils and industry. It also includes the Nansen International Environmental and Remote Sensing Centre, which was opened in 1992 in St. Petersburg in order to foster research and cooperation. Its mission is to make a scientific contribution to the observations, the understanding and the predictions of Global Change and Environmental Processes in the high North.[6]

The activities of NERSC include research and demonstration projects, a Ph.D. and post-doc programme, an international scientific exchange programme, specialized database products, and an outreach, science and technology marketing and conference programme. Its funding comes through three main avenues:

- core funding from co-founders
- competitive national and international funding, both public and private
- endowments, donations and sponsorship

NERSC's research strategy is to integrate the use of remote sensing and field observations with numerical modelling through the use of advanced data assimilation techniques. It has been highly successful and in 2005 won the EU's research prize, the Descartes Prize, which was awarded to the project 'Climate and environmental change in the Arctic'. The Nansen International Centre in St. Petersburg was one of

---

6   Nansen International Environmental and Remote Sensing Centre. 'About NERSC'. http://www.nersc.no/main/index2.php?display=aboutsummary [accessed 28 July 2008].

the main contributing institutions in this project. NERSC also seeks to commercialize its research through the companies Terra Orbit A/S and COTO A/S.[7]

Sweden

There have previously been several cooperative projects on renewable energy between various Swedish government ministries and Russia in the field of renewable energy. However, we have not been able to identify any currently ongoing bilateral projects. This reflects a change in the funding structure, with funding instead being channelled through Nordic multilateral agencies. Like Denmark, Sweden has focused on implementing projects through the Kyoto Protocol mechanisms, such as Joint Implementation (JI) projects.[8]

Royal Swedish Academy of Sciences

The Royal Swedish Academy of Sciences provides funding for international research collaboration, mostly through the participation of Russian researchers in Swedish projects, and exchange programmes. The academy has a significant focus on environmental issues in Central and Eastern Europe, where renewable energy can play an important role in reducing pollution and environmental degradation.

Iceland

Iceland is in the process of initiating a particularly interesting bilateral project: the creation of the Russian–Icelandic Institute of Renewable Energy. This is done under the auspices of the School for Renewable Energy Science (RES) in Iceland, with the support of the Icelandic Ministry of Foreign Affairs and Rannis, the Icelandic centre for research and research funding. This is still at the start-up stage, but there has been significant Russian interest in the initiative.[9] Discussions continue about the priority areas of the new institute, but the main focus is likely to be on education and research. It is also expected that there will be co-financing from Russia.[10]

Additionally, the School for Renewable Energy Science in May 2008 signed a partnership agreement with the prestigious MGIMO University of Moscow to facilitate cooperation in renewable energy. This will include partnership in education, exchange and research.

---

7   Nansen Environmental and Remote Sensing Centre homepage. http://www.nersc.no/main/index2.php [accessed 28 July 2008].

8   Interview with Gudrun Knutsson, Swedish Energy Agency, 5 August 2008.

9   Interview with Björn Gunnarsson, Academic Director of RES, 5 August 2008.

10   Interview with Björn Gunnarsson 2008.

Finland

The Ministry of Employment and Economy

The Finnish Ministry of Employment and Economy has a large number of ongoing bilateral cooperation projects with Russia in the field of renewable energy, reflecting the geographical proximity of the two countries. For the most part these are run by the Ministry of Employment and the Economy, but with the Ministry of Foreign Affairs responsible for overall coordination and supervision.

This funding takes the form of grants aimed at improving energy efficiency, establishing centres of technology and competence, education and training initiatives, feasibility studies and other technical assistance. In 2008, the total contribution to bilateral cooperation projects will be EUR 19.4 million, of which renewable energy and energy efficiency projects receive a considerable share.

Academy of Finland

The Academy of Finland mostly provides funding for international research collaboration through the participation of Russian researchers in Finnish projects, and exchange programmes. In this sense it is similar to the other Nordic research councils/academies of science.

Multilateral actors

There is a growing tendency for the Nordic countries to work through pan-Nordic or pan-Baltic organizations in their cooperation with Russia in the area of renewable energy. There is a wealth of actors, and their activities are significantly interwoven–with many institutions having the same focal areas and collaborating on the same projects, whether through technical expertise or co-financing. We start with one of the most important multilateral actors, the Nordic Council of Ministers.

The Nordic Council of Ministers

The Nordic Council of Ministers (NCM) has undertaken a range of programmes aimed at fostering cooperation in renewable energy and related areas. These programmes have a primary geographical focus on Northwestern Russia and include scholarship programmes, exchange programmes, the development of knowledge-sharing networks and various development programmes. In this section we highlight some key areas of operation of the Nordic Council of Ministers, before proceeding to look at various organizations that function under the auspices of the NCM.

One of the most visible NCM initiatives is the presence of a chain of information offices throughout Russia: in St. Petersburg, Arkhangelsk, Murmansk and Petrozavodsk. They seek to:

1. promote the interests and international cooperation of the Nordic Countries in Northwestern Russia;
2. enhance knowledge-building and networking between Nordic and Russian partner organizations;
3. coordinate the administration and implementation of the activities and programmes of NCM in Northwestern Russia.

The other NCM initiative most relevant to this book is Energy Experts Exchange programme, intended to foster collaboration and research between Russian and Nordic scientists.

Although the Nordic Council of Ministers does outsource many of its renewables projects to sub-organizations working under the NCM, such as Nordic Energy Research, it also undertakes its own projects in the areas of renewable energy, climate change and energy efficiency. A list of some of these projects is presented below. Finally, as noted above, there is significant overlap in many of the activities undertaken by various multilateral institutions. Some of these projects, while financed by the NCM, are carried out under the auspices of other multilateral institutions such as NEFCO. Here one should also be aware of the difference between sub-organizations within the NCM structure and other multilateral organizations that are entirely separate but also work together closely.

For Russian actors, this organizational tangle can be confusing. First they have to understand the concept of the 'Nordic countries' as a geographical unit, which does not make immediate sense to all Russians. Then they have to recognize that the countries comprising this group have their own multilateral organizations (like NCM), and that although the multilateral cooperation may be very loose compared to the EU, it is nonetheless significant. Finally, they have to understand that there are several parallel Nordic organizations, each with its own relatively independent sub-organizations. Only then can they start making sense of the opportunities for funding and cooperation.

Due to the vast number of relevant projects under NCM, we have categorized them in Table 6.1.

Here we present several of NCM's sub-organizations working with various facets of renewable energy – NordForsk, the Norwegian Innovation Centre, the Nordic Investment Bank and Nordic Energy Research.

Table 6.1   NCM projects in Russia on renewable energy/energy efficiency

Mechanism	Time frame	Title and/or project description	Location	Cost (Euros)
Scholarships	2006–2007	Nordic scholarship programme for energy experts. Provides young energy experts with an opportunity to gain further education, training and experience	Kaliningrad	134,228
Grant	2004–2006	Bioenergy as environmental factor and landscape resource. Focus on sustainable and enhanced bioenergy production and forest management. Includes integrating bioenergy with other renewable energy sources	Kaliningrad	112,752 Other partners €60,402
Grant	2004–2006	Bioenergy as an environmental factor in the Nordic–Baltic–Northwest Russian region. Project focuses on sustainable and enhanced bioenergy production and forest management.	Nordic, Baltic, Northwest Russia	132,700
Revolving credits	1997–2010	Energy efficiency revolving credit frame for St. Petersburg, Vodokanal. Aims to improve energy efficiency, improve level of services and reduce capital costs	St. Petersburg	150,000–500,000
Revolving credits/loans	2001–2010	Municipal water, sewage, energy and waste project in Novgorod. Aims to improve energy efficiency, level of services and reduce capital costs	Novgorod	150,000–500,000
Grant	2004–2006	Nordic Environmental Youth Camp 2005–2006. Goal was to enhance and deepen cooperation between Nordic environmental youth and NGOs and to include Russian and Baltic representatives. Focus on environmental issues and climate change.		20,000–150,000
Revolving credits/loans	2001–2010	NEFCO Cleaner Production Revolving Facility. Provides loans for small scale projects that reduce risk to humans and the environment. Includes energy conservation and the reduction of emissions.	Arctic region	>1,500,000
Revolving credits and loans	2005–2015	NEFCO Energy Saving Credits Revolving Facility. Investments in energy-saving measures in schools, hospitals, etc. or investments in fuel switch from heavy oil/gas to bio-fuel in local boiler houses	Arctic region	10,000,000 Local contribution €1 million
Revolving funds	1997–2006	NEFCO Energy Savings Programme. Investments in energy saving measures in schools, kindergartens and hospitals in Northwest Russia. Aims to achieve considerable energy savings, reduce consumption of fossil fuels – especially heavy oil. Includes 30 approved projects.	Arctic region	4,500,000
Grant	2006–2009	The Arctic Hydrological Cycle Monitoring Modelling and Assessment Programme, Arctic-HYDRA	Arctic region	659,425
Grant	2003–2006	The Pasvik Programme – Development and implementation of an integrated environmental monitoring and assessment programme in the joint Finnish/Norwegian/Russian border area	Arctic region	1,605,574
Grant	2005–2006	Nordic Environmental Youth Camp 2005–2006. The goal was to enhance and deepen cooperation between Nordic environmental youths NGOs and also to include Baltic and Russian representatives in the cooperation. Focus on environmental issues such as climate change.		20,000–150,000
Grant	2002–2006	Exchange of Civil Servants. Aims to build networks between civil structures of Northwest Russia and the Baltic states and civil structures in Nordic countries		150,000–500,000

NordForsk

The Nordic Research Board (NordForsk) is a sub-organization of the Nordic Council of Ministers dealing with the full range of Nordic research collaboration. It has supported significant Russian–Nordic collaboration in several areas, but little concerning renewable energy. One reason is the structure of NordForsk. It sponsors various centres of excellence dealing with various research areas; however, none of these have dealt with areas related to renewable energy.[11] Thus there are no top–down projects dealing with renewable energy, only bottom–up projects initiated by researchers within open calls.

The nature of funding has also been affected by bureaucratic and policy changes within NordForsk. Up until the spring of 2007 Northwestern Russia was included in the NordForsk definition of 'neighbouring countries', one of its areas of operation. However, in 2007 Russia became a separate area of operation, and that has made funding more difficult. NordForsk has taken several steps to rectify this problem, including holding a symposium in St. Petersburg in 2007 to foster interest. In spring 2008 it launched a call asking its centres of excellence if they were interested in acquiring Russian partners.[12] Cooperation has also been limited due to a lack of co-financing from the Russian side; however, it is hoped that this will change in the coming years.

Nordic Innovation Centre (NIC)

The Nordic Innovation Centre is the Nordic Council of Ministers' key instrument for promoting an innovative and knowledge-intensive Nordic business sector. At present, the Nordic Innovation Centre is not particularly active in renewable energy cooperation projects in Russia, but it is working to increase its focus on this area.

The NIC hopes to finalize an initiative in the course of 2009 called Clean-Tech, which it will operate together with Nordic Energy Reseach and a range of other Nordic actors dealing with renewable energy.[13] It is hoped that this will enable the Nordic Innovation Centre to put a more significant focus on renewable energy projects and cooperation with Russia. This initiative will utilize a bottom–up funding mechanism: interested actors will submit an Expression of Interest on the NIC website, whereafter the NIC will investigate the feasibility of the project. If the NIC finds the project relevant, it will offer a grant to finance up to 50 per cent of the project. The NIC usually focuses on projects with a timespan of between one and three years.[14]

---

11   Interview with Harry Zilliacus, Senior Advisor for NordForsk, 1 August 2008.
12   Interview with Harry Zilliacus 2008.
13   Interview with Sigridur Thormodsdottir, Senior advisor NIC, 4 August 2008.
14   Interview with Sigridur Thormodsdottir 2008.

Nordic Investment Bank (NIB)

In the neighbouring areas of the member countries, NIB grants loans to projects that support economic development and to investments to improve the environment. In February 2008 the NIB announced that it will provide EUR 1.5 billion in loans to cut hazardous emissions in the Baltic Sea and mitigate climate changes.[15] NIB will allocate these funds through two new environmental lending facilities – the Baltic Sea Environment Financing Facility with a framework of up to EUR 500 million, and the Climate Change Mitigation and Adaptation Energy Facility with a framework of up to EUR 1 billion. Funds for the new facilities will be earmarked within the Bank's ongoing lending activities.

The environmental investment loans are granted to governments, governmental authorities, institutions and companies. They have long maturities, which reflect the projects' cash flow and lifetime.[16] Additionally, NIB's provision of credit is well suited for investments that secure energy supplies, improve infrastructure or support research and development. The Climate Change Mitigation and Adaptation Energy Facility will focus on financing projects within renewable energy (hydro, wind, biomass, geothermal and solar power) and the more effective use of energy, as well as projects using cleaner production technologies that reduce greenhouse gas emissions in industries. The facility will also target projects dealing with the adaptation of power networks and infrastructure to climate change, such as extreme weather conditions.[17]

Nordic Energy Research (NER)

Nordic Energy Research is the Nordic Council of Minister's sub-organization dealing with energy issues. It was established as an institution in 1999, as a continuation of the Nordic Energy Research Programme, which had existed since 1985. NER aims primarily to support research and development activities through grants, mobility support, network and project funding and support to seminar and course activities.[18] NER has shown interest in the potential of the Russian market, and is active in several cooperation projects.

Renewable energy is one of NER's priority fields, and as such it has been active in several areas, including creating Centres of Excellence to pool expertise

---

15   Nordic Investment Bank. 'News'. http://www.nib.int/newsen/1211961178.html [accessed 3 August 2008].

16   Nordic Investment Bank. 'Lending'. http://www.nib.int/lending/neighbouring. html [accessed 3 August 2008].

17   Nordic Investment Bank homepage. http://www.nib.int/home [accessed 23 June 2009].

18   Nordic Energy Research. *Strategic Action Plan 2007–2010*. Oslo: Nordic Energy Research, 2006, p. 4.

Table 6.2   The three types of projects implemented by NER[19]

	1. Capacity- and competence-building projects	2. Business development and innovation projects	3. Integrated projects
Quality perception	Scientific / internal	External quality / user-driven problem	Internal and external (Scientific/user-driven problem)
Nordic relevance	Relevance to the Nordic energy sector and/or energy policy	Relevance to the Nordic energy sector and industry	Relevance to the Nordic energy sector and industry
Number	Min. 4 Nordic countries and autonomous areas. Baltic countries and Northwest Russia appreciated	Min. 3 Nordic countries and autonomous areas. Baltic countries and Northwest Russia appreciated	Min. 3 Nordic countries and autonomous areas. Baltic countries and Northwest Russia appreciated
Actors	Project partners mainly from academia, but users from industry, energy policy and energy sector highly appreciated in advisory bodies	Projects partners from research institutes, industry, energy sector, etc.	Projects partners from academia, research institutions, industry, energy sector, etc.
NER finances	Up to 85% of the total project eligible costs	Up to 50% of the total project eligible costs	Up to 75% of total project eligible costs
NER allocates	Max. 3 mill NOK per year	Max. 3 mill NOK per year	Max. 3 mill. NOK per year
Project length	Max. 4 years	Max. 2 years	Max. 4 years

and increase the focus on priority areas. Several of these Centres of Excellence have involved Russian participation:

- the Nordic Centre of Excellence in Photovoltaics (PV)
- Distributed Generation Integration in the Nordic energy market (DIGINN)
- the Nordic Centre of Excellence on Hydrogen Storage Materials.

In addition, Nordic Energy Research emphasizes cooperation through capacity- and competence-building projects such as Ph.D. and post-doc grants, visiting scholar grants, research courses and summer schools, workshops, courses and seminars. Other activities include mobility grants for Ph.D. candidates and young researchers to other Nordic or international institutions, and the holding of international scientific conferences. In 2007 Nordic Energy Research launched the programme NORIA-energy. The aim here is to strengthen the Nordic research and innovation area in new energy technologies and systems, and to aid Nordic decision-makers in developing efficient policies on science, technology and investment in new energy technologies and systems.[20]

---

19   Source of data used to compile figure: Nordic Energy Research 2006, p. 12.

20   Nordic Energy Research. *Annual Report 2007.* Oslo: Nordic Energy Research, 2008, p. 17.

With regard to funding structure, NER has several bottom-up projects, reflecting its aim to be 'at the intersection between the authorities, research and industry'.[21] Its primary funding mechanism is grants. However, projects require co-financing, with the level depending on the type of project. This is highlighted clearly in Table 6.2, which provides an overview of the three types of projects implemented by NER.

In terms of financing, NER will have a total financing of NOK 185 million for the period 2007–2010. National contributions amount to NOK 110 million. From the Nordic Council of Ministers NOK 12 million is expected, and NOK 8 million from other projects (EU projects, administration projects, etc.) whilst approximately 55 million NOK is expected as co-financing. The total expenses are distributed in the following way: NOK 90 million is allocated to capacity and competence projects, innovation projects and integrated projects. NOK 45 million is expected in co-financing. NOK 10 million is allocated to policy studies, and a similar amount is expected in co-financing. Expenses for other projects are NOK 8 million. Information, coordination and administration total NOK 22 million, or 5.5 million per year (12 per cent of total expenses).[22]

BASREC

The Baltic Sea Region Energy Cooperation (BASREC) functions under the Council of the Baltic Sea States (CBSS). In addition to the Nordic countries and Russia, it includes Estonia, Latvia, Lithuania, Germany and Poland. BASREC cooperation has provided the CBSS member countries with a forum to establish a regional perspective on energy policy strategies. The networks and BASREC´s organizational structure give organizations and business actors in the energy sector a natural base for analysing the possibilities to develop the market framework and rules in order to make energy supply more efficient and to reduce environmentally problematic impacts of energy production, use and transmission.[23]

Since the end of 2005, BASREC's funding has been cut significantly, and several subsidiary structures have been discontinued. This has also resulted in a lack of funding for cooperation projects.[24] There is, however, a draft communiqué circulating that proposes the establishment of a pool of funds to be used for various initiatives. This communiqué was discussed at the next ministerial meeting in Copenhagen in September 2008.[25]

---

21   Nordic Energy Research 2006, p. 5.
22   Nordic Energy Research 2006, p. 26.
23   BASREC. 'Baltic Sea Region Energy Co-operation'. http://www.cbss.org/Energy/baltic-sea-region-energy-cooperation [accessed 21 June 2009].
24   Interview with Phillip Saprykin, BASREC Sectretariat, 4 August 2008.
25   Interview with Phillip Saprykin 2008.

NEFCO

The Nordic Environment Finance Corporation (NEFCO) is an international finance institution established in 1990 by the five Nordic countries. NEFCO has financed a wide range of projects in Russia and other East European countries aimed at achieving cost-effective environmental benefits.

NEFCO's portfolio currently comprises nearly 300 small and medium-sized projects spread across various sectors, including chemicals, minerals, metals, food, engineering, agriculture, water treatment, power utilities, municipal services, waste management, nuclear remediation, environmental management and environmental equipment manufacturing.[26]

Emphasis is placed on direct investments from, for example, public–private partnerships and corporate public services. The structure of any project supported by NEFCO needs to create a reasonable equilibrium between risk and reward for all interested parties. The aim is to achieve a fair and transparent balance between the structuring of investments and their environmental return. Viewed in the context of associated risk, NEFCO aims to offer competitive terms at all times.[27]

NEFCO has a range of funding mechanisms available, which will be highlighted below. Moreover, through its network of partnerships, NEFCO supplements finance from other interested parties and financial institutions. NEFCO also collaborates with bilateral environmental assistance programmes.

NEFCO Investment Fund

The NEFCO Investment Fund amounts to approximately EUR 114 million. It offer loans and equity financing. In some cases subordinated loans and loans with equity features can also be provided. The loans are from medium to long term, and are provided at market conditions.

Nordic Environmental Development Fund (NMF)

Through this fund, originally established by the Nordic Ministers of Environment in 1995, NEFCO seeks to support the realization of projects that otherwise would not materialize or could be realized only later in the future. Local participation in the financing is required. Contributions from the fund can be provided as grants for the procurement of goods or services (cash subsidies) or to reduce the borrower's debt-service costs. The maximum grant is one-third of the total project cost. The

---

26　Nordic Environment Finance Corporation (NEFCO). 'Introduction'. http://www.nefco.org/introduction [accessed 31 July 2008].

27　NEFCO 'Introduction'.

capacity of the fund is approximately DKK 300 million. The NMF also includes the Revolving Facility for Cleaner Production investments, which provides loans directly to enterprises implementing cleaner production programmes, and the Energy Savings Programme and Energy Savings Credit Facility, which supports a range of small energy-efficiency investments.[28]

Environmental hot spots in the Barents region

NEFCO has a special mandate to work with environmental issues and projects in the Arctic and the Barents regions. One of the main financial tools for doing this is the Barents Hot Spots Facility (BHSF), managed by NEFCO on behalf of the governments of Finland, Iceland, Norway and Sweden.

Carbon finance and funds

The Baltic Sea Region Testing Ground Facility (TGF) was established at the end of December 2003 by the governments of Denmark, Finland, Iceland, Norway and Sweden. It is a EUR 35 million regional carbon finance facility which purchases carbon credits under the Kyoto Protocol: Assigned Amount Units (AAUs), and Emission Reduction Units (ERUs) from energy-related and other projects on behalf of its investors. It is structured as a Public–Private Partnership (PPP) between governments, private sector utilities and industrial companies in the Baltic Sea region. The TGF is managed as a trust fund by NEFCO.[29]

NEFCO Carbon Fund (NeCF)

The NEFCO Carbon Fund (NeCF) is a global carbon fund based on a Public–Private Partnership model, launched in March 2008. Vested in the form of a trust fund administered by the Nordic Environment Finance Corporation, it is an instrument for purchasing greenhouse gas emission reductions under the joint implementation (JI) and clean development mechanisms (CDM).[30] The NeCF will invest in a wide range of projects by providing financing for renewable energy, energy efficiency, fuel switching and other investments, in many countries, Russia among them.

Arctic Council Project Support Instrument (PSI)

In March 2005, the Arctic Council established the Project Support Instrument, a financial initiative that focuses on actions to prevent pollution of the Arctic.

---

28   Nordic Environment Finance Corporation (NEFCO). 'Financing'. http://www.nefco.org/financing [accessed 31 July 2008].

29   Nordic Environment Finance Corporation (NEFCO). 'Carbon Finance'. http://www.nefco.org/financing/carbon_finance [accessed 31 July 2008].

30   NEFCO 'Financing'.

NEFCO was appointed as the Fund Manager. The PSI is a mechanism for financing specific priority projects already approved by the Arctic Council. The intention is to invite interested Arctic Council member states, observers and others to pledge contributions to the PSI.[31]

Project-specific funds

NEFCO administers several special-purpose funds on behalf of various donors, for the specific support of certain projects. Donors here include the Nordic countries, the Dutch government, the Nordic Council of Ministers, EU Phare (the enlargement assistance programme), the Swedish Energy Agency and the Global Environmental Fund (GEF) through the Helsinki Commission (HELCOM). NEFCO has for these project-specific funds been assigned a total of EUR 83.3 million, of which EUR 23.4 million represent completed funds.[32]

Barents Euro-Arctic Council

The members of the Barents Euro-Arctic Council are Denmark, Finland, Iceland, Norway, Russia, Sweden and the European Commission. Cooperation was launched in 1993 on two levels: intergovernmental (Barents Euro-Arctic Council, BEAC), and interregional (Barents Regional Council, BRC), with sustainable development as the overall objective. There are several joint working groups working with priority areas; the Energy Working Group and the Working Group on Environment are of particular relevance in terms of renewable energy.

The Energy Working Group emphasizes energy efficiency as a means to increase economic performance and reduce the environmental load. Centres have been established in Murmansk, Kirovsk, Petrozavodsk, Arkhangelsk, Syktyvkar and NaryanMar, to services for local authorities, utilities and industry. The entry into force of the Kyoto Protocol has paved the way for Joint Implementation projects in the Barents region. Other activities include the promotion of bioenergy, for which there is a huge potential in the region.[33]

The Working Group on Environment pays particular attention to cleaner production, the elimination of environmental 'hot spots' in the Russian part of the Barents region, and the conservation of biological diversity and sustainable forestry. A special project preparation fund (EUR 3 million) for environmental 'hot spots' has been created with NEFCO as fund manager. Russian financing is

31   NEFCO 'Financing'.
32   NEFCO 'Financing'.
33   Barents Euro-Arctic Region. 'Information Document', 2005. http://www.barentsinfo.fi/beac/docs/11675_doc_CSO.2005.18ENGBarentsInfosustainabledevlp.pdf [accessed 20 June 2009)], p. 2.

also expected to be made available. There is a commitment to eliminate the 'hot spots' by 2013.[34]

The lack of funding is an obstacle to sustaining and increasing cooperation. Most funding is national, with Interreg (and Tacis) as important additional sources. The region's eligibility for funding under the European Neighbourhood and Partnership Instrument (ENPI) after 2007 should be ensured. The Nordic Council of Ministers' new Russia Programme opens up opportunities for greater interaction with the Barents cooperation.[35]

Bellona Foundation

The Bellona Foundation is an independent non-profit foundation based in Norway and with offices in Murmansk, St. Petersburg and Moscow. They work with various energy issues, with a significant focus on the development of renewable energy in Russia. To this end they have established the Northwest Russia Renewable Energy Forum as an arena for information exchange, capacity-building, research and innovation, technology and product transfer, as well as investment in renewable energy. It seeks to disseminate information about the potential for renewable energy on the Kola Peninsula and to remove obstacles (psychological, bureaucratic, economic, technological, and juridical) to implementation of renewable energy in the region.[36] It should be noted that the Bellona Foundation focuses on information dissemination, and bringing relevant actors together, and as such there is no large-scale project implementation.

Northern Dimension Environmental Partnership (NDEP)

The NDEP embodies an innovative approach to delivering effective solutions to some of the most pressing environmental problems facing Northwest Russia, which, to date, consists of a EUR 2.4 billion pipeline of projects. As of the end of 2006, the NDEP had leveraged investments of more than EUR 1.5 billion for environmental projects.[37]

The NDEP's environmental programme consists of 15 priority projects whose combined cost exceeds EUR 2 billion. These projects aim to deliver environmental

---

34   Barents Euro-Arctic Region. 'Cooperation in the Barents EuroArctic Region. Target: Stability and Sustainable Development', 1 June 2005. http://www.barentsinfo.fi/beac/docs/11675_doc_CSO.2005.18ENGBarentsInfosustainabledevlp.pdf [accessed 20 June 2009], p. 3.

35   Barents Euro-Arctic Region 2005, p. 4.

36   Bellona. Title Page. http://www.bellona.org/ [accessed 23 June 2009].

37   Northern Dimension Environmental Partnership (NDEP). 'Background'. http://www.ndep.org/home.asp?type=nh&pageid=5#how2 [accessed 30 July 2008].

solutions to the Northwest of Russia in the areas of district heating, solid waste management, wastewater treatment and energy efficiency.[38]

The NDEP, which began operating in 2001, pools the joint resources of the international financial institutions (IFIs) operating in the region – European Bank for Reconstruction and Development (EBRD), European Investment Bank (EIB), Nordic Investment Bank (NIB) and the World Bank. Through the NDEP Steering Group, the NDEP draws on the collective expertise and know-how of the IFIs, the European Commission and the Russian Federation. Every project has an assigned IFI acting as Lead Implementing Agency, which means that all projects can follow a well-planned and prioritized programme including support for sector reform and institution building. In addition, through its involvement in the NDEP, the EIB has a lending mandate for financing environmental projects in Russia for the first time.[39]

The main funding mechanism of the NDEP is the NDEP Support Fund, which receives financial contributions from donor governments, including Russia, to leverage environmental investment in the Northern Dimension area. The purpose of the Fund is to catalyse investment by mobilizing grant co-financing for projects prioritized jointly by the NDEP Steering Group and the Nuclear Operating Committee and approved by the NDEP Assembly of Contributors.

The NDEP Support Fund catalyses environmental investments by mobilizing grant co-financing for leveraging loans from the IFIs. These grant funds have a multiplicator effect on environmental investment in the area as they help to secure larger IFI loans to finance the major share of the investments. Without a support fund, the NDEP would not be able to gain the necessary strength. The NDEP Support Fund has two key areas: nuclear safety and environmental projects.[40]

Case study: The Baltic Billion Fund II

The Baltic Billion Fund II was a business-sector promotion programme focusing on Swedish companies in the Baltic countries and northwestern Russia. At the time of the evaluation SEK 824 million of the allocated 926 million of funding had been spent. Of this 824 million, 748 million was used on projects, and 76 million on administration.

38　Northern Dimension Environmental Partnership (NDEP). 'Project Pipeline'. http://www.ndep.org/projects.asp?type=nh&cont=prjh&pageid=15&content=projectlist [accessed 20 July 2008].

39　NDEP 'Background'.

40　Northern Dimension Environmental Partnership (NDEP). 'The Structure of NDEP'. http://www.ndep.org/home.asp?type=nh&pageid=6 [accessed 30 July 2008].

There are some interesting lessons to be learned from the Baltic Billion Fund II, especially concerning problems of measuring outcomes. Although trade increased greatly between Sweden and the Baltic Sea area, the connection between company participation in the programme and outcomes is weak, and in many cases impossible to establish with scientific methods.[41] Indeed many of the business deals concluded during the programme timeframe might have occurred anyway without support, thus it is conceivable that the business value was overstated.[42]

There was a clear connection between the emphasis of the projects and achieved business outcome. The projects that showed the greatest business value were characterized by being aimed at individual companies and defined activities, such as establishment in the market or a specified business deal. Projects emphasizing more general activities, such as information, training and establishing contacts, seldom had much demonstrable business value. On the other hand, this may also be due to the problem of measuring the outcome of such projects.

According to the evaluations that the Swedish Agency for Public Management has scrutinized, many of the contributions have contributed to positive development of the participating companies, through increased knowledge, increased insight regarding market potential, increased interest in the markets and increased ability to conduct business in the programme area. However, assessments of the quality and relevance of content for the companies are lacking. It is therefore difficult to make a collective assessment of the importance of the programme for development of the companies on the basis of the evaluations.[43]

The evaluations of the programme indicate that the horizontal goals (like equality and sustainable development) have had limited effect on the programme. Only a few evaluations brought up these goals at all. Judging from the evaluations, the different horizontal goals appear to have been taken into account to varying extents. The environmentally-oriented goals showed the greatest effect. By contrast, the set goals regarding equality, sustainable social development and democracy appear, on the basis of the evaluations, neither to have been taken into account nor achieved.[44]

The set goals of Baltic Billion Fund II were not quantified. Therefore, the final evaluators were unable to say whether the outcome was sufficient to conclude that

41   Swedish Agency for Public Management. *Baltic Billion Fund 2: A final assessment.* Stockholm, 2006. http://www.statskontoret.se/upload/Publikationer/2006/200608_english summary.pdf [accessed 21 June 2009], p. 2.

42   Swedish Agency for Public Management 2006, p. 3.

43   Swedish Agency for Public Management 2006, p. 3.

44   Swedish Agency for Public Management 2006, p. 3.

the goals had been fulfilled. All the same, the assessment was that goal fulfilment of the programme was low.[45]

The final evaluator's most important argument with regard to this is that development would have been roughly the same, even without Baltic Billion Fund II. The value of trade has increased strongly, and each year thousands of Swedish companies do business on the markets in the programme area without public support. In relation to this, the outcome of Baltic Billion Fund II stands out as limited, as regards the number of participating companies and the business value attained. There is therefore a considerable risk that the deadweight effects in the programme have been significant: in other words, that support has been given unnecessarily.[46]

There are only a few projects and a handful of companies that can show real business results from the programme. Often these are cases where major multinational companies have concluded big business deals connected to international development projects.

The key factors for companies to succeed in the new markets were perseverance, motivation, experience and resources. If these are lacking in the companies, there is little to indicate that an internationalization process would succeed. It is difficult to formulate support to influence these conditions effectively.[47]

It is not possible to determine whether the programme had any effect in the sense of a greater number of companies doing business in the programme area. For example, considerable resources were put into projects directed towards Russia. During the programme period, the value of trade with Russia increased, but the number of companies trading with Russia decreased.

Additionally, Baltic Billion Fund II is still affected by the fact that the Swedish government lacks the administrative infrastructure required for running operative systems of this type and scale. Above all, there is lack of administrative IT support for financial accounting for the programme. This means that, to a great extent, the financial accounting is dependent on manual routines.[48]

Lessons learned

We will now look at some of the lessons learned from various actors about their experiences in funding projects and cooperation in Russia. Many of these are similar to the experiences highlighted elsewhere in this book.

---

45   Swedish Agency for Public Management 2006, p. 4.
46   Swedish Agency for Public Management 2006, p. 4.
47   Swedish Agency for Public Management 2006, p. 4.
48   Swedish Agency for Public Management 2006, p. 4.

A range of institutions have encountered significant problems with the bureaucratic hierarchy of the Russian system. Contacting researchers directly can often be difficult, as one often needs to speak to their superiors first, and they are not necessarily as interested in entering into cooperation. Additionally, finding relevant researchers has also presented a significant problem. Various actors have also encountered scepticism on the part of Russian scientists when it comes to research focusing on sustainable energy due to the scepticism prevailing among many Russian scientists as to the anthropogenic causes of climate change.

Other problems with collaboration are due to communication difficulties, and cultural differences. Indeed, from the lessons learned from the case study of the Baltic Billion Fund it could be said that the key factors for success in the new markets are 'perseverance, motivation, experience and resources'. A crucial point is to be aware of the cultural and bureaucratic differences, and to be patient.

Cooperation has also been limited due to a lack of co-financing from the Russian state. One reason for this could be the Nordic bias of Nordic projects: the Russians might perceive these as being overly Nordic, and thus not see it in their interest to participate with funding in such collaborative projects.

At the same time, this also reflects a lack of interest in renewable energy on the part of Russia's federal government. Here it should be noted that the government is more than willing to allow Nordic actors to keep undertaking energy efficiency projects, as long as most funding comes from Nordic actors. In terms of energy, Russia's main priorities remain oil and gas, even though increased production of renewable energy sources would allow them to export more, thus contributing significantly to budget revenues.

In terms of funding mechanisms it was hard to identify any clear lessons learned, despite the wide variety of funding mechanisms used. The main reason is the lack of evaluation of funding mechanisms and their effectiveness in terms of various projects. There is a distinct lack of critical analysis of funding mechanisms, and further work should be undertaken in this area.

Summing up the Nordic scene

There is wide range of Nordic actors working with research and innovation on renewable energy, environmental studies, hydrogen, climate change and energy efficiency, and most of them have some interface with Russia. They form a complex interwoven web, sharing focal areas and often collaborating and financing the same projects. The nature of this web and the shared priorities make it difficult to pinpoint which actors have the best expertise in particular areas. This would indicate a significant amount of bureaucratic baggage, and should provide an incentive to streamline funding and operating structures.

At the same time there is much potential for private partnerships and working with NGOs. We have not listed private Nordic actors involved in renewable energy in Russia due to their sheer number and the constraints of our time: we estimate there to be over 200. The extent of this penetration offers interesting potential for further research. A first step might be to catalogue these actors.

In addition to technology companies, there are several Nordic consultancies that offer advice on renewable energy solutions and entering the Russian market. These have been supplemented by various NGOs which aim to foster cooperation, and highlight key issues, such as the importance of environmental awareness.

One thing that became clear in research this chapter was the shift among Nordic actors towards using the mechanisms provided under the Kyoto Protocol's flexible Joint Implementation (JI) to engage with Russia. Denmark and Sweden in particular have been representative of this tendency. With the vast amount of emissions emanating from Russia, these mechanisms provide an effective instrument for implementing renewable energy projects, and possibly cutting down on bureaucratic baggage. Of the 163 JI projects currently underway, 109 are located in Russia.[49]

Another interesting aspect of the involvement of Nordic organizations in Russia is their focus on Northwestern Russia. Although definitions of precisely what constitutes this area vary from organization to organization and change over time, the picture is essentially consistent: Nordic attention is focused almost exclusively on Murmansk, Arkhangelsk, Karelia, St. Petersburg or Kaliningrad. That means that Moscow, the most important city in Russia for both science and business, and many other important areas of the country (for example the impressive research institutes in Siberia) do not figure at all on the Nordic horizon. Since Russia borders on more countries than any other country in the world, and many of those borders are far from Moscow and other Russian centres of renewable energy, other countries may also have something to learn from this weakness in the Nordic engagement with Russia. We believe that there is also much to learn from the model we present in the next section which maps the comparative advantages of the five Nordic countries and Russia to seek out complementarities. This is a useful exercise for any actor desiring to establish renewable energy cooperation with Russia.

The Nordic countries and Russia: Comparative advantages and complementarities

This section outlines some of the areas within renewable energy in which Russia has particular strengths or weaknesses and analyses how these match the foci

---

49   UNEP/RISØ. 'JI Projects'. http://cdmpipeline.org/ji-projects.htm [accessed 14 August 2008].

of the Nordic countries. The five Nordic countries represent a microcosm of the renewable energy world, with each country having different natural resources and excelling in different technological areas (Denmark – wind; Iceland – geothermal; Sweden hydro and bioenergy, etc.). It should have something to offer for readers from other parts of the world, and actors seeking to engage in renewable energy cooperation with Russia can benefit from undertaking a similar exercise. Our analysis is based on the over 100 interviews listed in the appendix.

Social and natural sciences

The Russian expertise and literature tend to focus narrowly either on the technologies for renewable energy or on the technical or theoretical potential of various renewable energy sources and their distribution across the country.

There are two reasons for this situation. Firstly, the field is strongly anchored in engineering, and many of the key actors who actively brand themselves as renewables experts have engineering or similar backgrounds. Secondly, it reflects a Soviet tradition of producing endless amounts of facts and numbers that have as little political relevance as possible, because during the Soviet period this was the safest thing to do and allowed the author to be entirely objective without getting into trouble with the Communist Party.

The result is that commercial, sociological and political approaches to renewable energy have had very low priority. This is also one reason why it is so difficult to find reliable information about the real market potential for renewable energy in Russia. It also strengthens the relative importance of fields where basic science plays an important role, for example hydrogen or solar power.

Another, and perhaps more important, consequence of this pattern concerns Russian–Nordic cooperation. The complementarities become obvious: strong Russian basic research in the natural sciences, strong Nordic skills in social science, innovation, commercialization and marketing.

Resources and competences

In judging a country's strengths and weaknesses on specific forms of renewable energy, one can take into account both (1) the natural resource base for that form of renewable energy (waves for wave power, wind for wind power, etc.); (2) the country's scientific strengths in the form of renewable energy and (3) the amount of investment in that form of renewable energy (both governmental and private investment). In Table 6.3 we have made a rough tripartite assessment of Russia's potential for various forms of renewable energy, taking into account scientific strengths, natural resources and investment.

Table 6.3 **Russian potential and Nordic overlap for various forms of renewable energy**

Renewable energy form	Russian scientific strength	Russian resource base	Russian investment	Nordic focus	Sum top joint areas
Hydrogen[50]	5	5	5	10	25
Solar	5	3	5	8	21
Bio-energy	3	4	2	10	19
Small hydro	2	4	2	9	17
Wind	2	2	3	10	17
Tidal	3	2	2	4	11
Wave	1	1	1	6	9

These assessments are combined with a simpler assessment of the foci of the Nordic countries. By adding these sums, we can arrive at a rough idea of some of the top joint Nordic–Russian areas within renewable energy. We stress that these are our personal, subjective assessments (drawing on the interviews) and other people may reach other conclusions. Nonetheless these assessments give a reasonably reliable overall picture.

Since Russia is in focus here, the Nordic strengths are only given a simple lump assessment. Russian strengths are rated 1–5, where 5 is best. Nordic strengths are rated 1–10, where 10 is best. The assessments in Table 6.3 build on the following considerations:

Hydrogen Russia's position in hydrogen partly grew out of its rocket science and car manufacturing traditions, and partly out of its strength in nano-science and materials for energy applications. One area where Russian milieus excel in particular is membranes for hydrogen fuel cells. As we shall see below, hydrogen has been the object of some unusually large and long-term investment.

Bio-energy Russia has a strong track record in biomass/combustion technology and materials for energy applications (e.g. energy storage). Even during the Soviet period, when the real cost of transportation was not factored in, some locations were considered too remote for the transportation of coal, natural gas or electricity. In such locations a tradition of using biomass was upheld. Comparatively speaking, however, this tradition is now waning since there has been rapid development in Western technology in recent years (pellets, bioreactors, second generation bio-energy and marine bio-energy) and this is a relatively applied science where Russia cannot float on old Soviet excellence in highly theoretical and mathematical fields

---

50   Hydrogen is not a source of renewable energy, but a potential carrier. Because of its prospective central role in an expanded renewables sector it is nonetheless included here.

to the same extent as in some other renewable energy areas. And although Russia has huge wood resources, these are mostly very remote, and there are significant problems with logistics. This means that bio-energy fits nicely into the niche of remote settlements in the High North discussed elsewhere in this book, but so far has less perspective for truly large-scale development.

Wind energy   As the world's largest country, Russia's territory inevitably includes some windy places. However, as a whole Russia is not comparable to the Nordic and other West European countries when it comes to wind power potential. Western Europe is located on the eastern shore of the North Atlantic, Russia is perhaps the world's most continental country and has much less wind on average. Although Russia does also have long shorelines in Arctic and the Far East, these are located very far away from population centres. Wind power therefore has comparatively less potential in Russia than in Nordic countries such as Denmark, Iceland or Norway. On the other hand, the long, windy, remote coastal areas in the North can be a good niche for the small-scale establishment of renewable energy, as discussed in the section on Russian markets for renewable energy in Chapter 2.

Solar power   Russian solar power technology supplied the Mir Space Station and its predecessors, the secret Almaz manned military space stations (1971–78), with electricity for several years, proving its reliability. These traditions of advanced photovoltaics for the Soviet and Russian space programmes laid the foundation for a strong tradition in basic and applied science in solar power. This tradition is upheld by a large number of institutes, and has also been translated into several clusters of commercial companies that produce photovoltaic materials and equipment for export. Finally, although much of Russia is sub-Arctic or Arctic, much of it is not. Contrary to popular perception, high temperatures in themselves are not good for photovoltaic solar power, in fact they decrease the efficiency of solar panels. And as the most deeply continental country in the world, many parts of Russia (notably most of Siberia) have very many sunny days.

Tidal energy   Again, as a huge country with most of its population concentrated in its southern and highly continental parts, Russia does not have the ideal coastline for tidal power. There are a few world class locations for the development of tidal power, notably on the Kola Peninsula and in the Far East. These are very attractive, but will never be a big part of Russia's energy supply.

Wave energy   For the same geographical reasons as noted under tidal energy, Russia's potential when it comes to wave power is limited.

Mapping the complementarities

The map opposite is an attempt at bringing out some geographical aspects of the complementarities between Nordic and Russian positions in renewable energy.

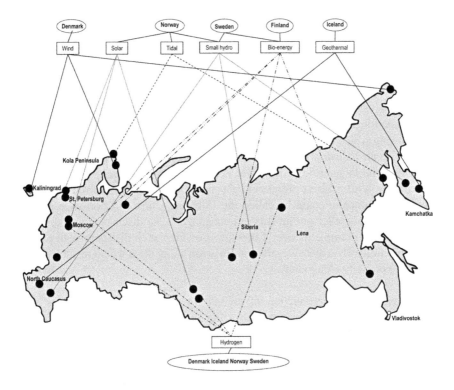

Figure 6.1   Nordic–Russian complementarities

The map shows where in Russia there is potential for the development of the various forms of renewable energy that each Nordic country specializes in.

The map above shows several things:

- Because of Russia's immense size – 17 million km^2 and 11 time zones – it has a diverse geography and a rich resource base that offer opportunities for each of the Nordic countries. Conversely, the Nordic countries can be interlocutors for Russia on almost any form of renewable energy.
- In practice each of the Nordic countries is mainly interested in cooperating within its own areas of specialization. The many potential linkages between Russia and the Nordic countries' different specializations within renewable energy therefore also pose a challenge to a coordinated Nordic approach to cooperation with Russia.
- The renewable-energy opportunities in Russia are distributed throughout the country.
- Most Nordic actors work exclusively with Northwestern Russia – which does not include Moscow, Central Russia, the South, Siberia or the Far

East. This precludes the majority of the interesting opportunities, since most of them are located outside Northwestern Russia.

The 'Siberian Curse': A business opportunity in disguise?

The Volga River is located in the European part of Russia, but most of the other big rivers are in Siberia: Irtysh, Kolyma, Lena, Ob, and Yenisei. These rivers were the arteries along which people and goods moved during the expansion of the Russian empire. For an industrial country it is however seen as a big disadvantage that so much of the large-scale hydropower is located in remote areas. This paradox of a country with massive resources, but many of which are too remote to be utilized economically could be called the 'Siberian Curse', to borrow a phrase from a recent bestseller on the Russian economy.[51] The distance from the Lena to Moscow, for example, is over 4,000 km. Covering such distances with power lines is prohibitively expensive, and transmission along them would result in large losses of electricity along the way.

However, for the production of silicon for solar power applications and hydrogen this remoteness is not such a big disadvantage. Their production is highly energy-intensive. Hydropower in particular is highly attractive for their production, because it makes it possible to market output as coming from a green energy production chain. In some ways the remoteness of Siberia's rivers could even be an advantage, since it can make large-scale hydropower very cheap due to lack of local demand. For the production of silicon, this is perfect, since the output is highly compact and easy to transport over long distances. For the production of hydrogen this will depend on the fruitfulness of the current efforts to develop cost-efficient technologies for hydrogen storage and transportation. On the other hand, if such technologies are not developed, the entire hydrogen economy will turn out to have been a pipe dream.

Norway's Renewable Energy Corporation (REC) recently decided to invest USD 1.2 billion in new silicon production facilities in Becancour, Quebec, after considering over 100 locations in 16 countries.[52] The decisive reason for choosing Becancour was the offer of a 20-year contract for power from Hydro Quebec with stable pricing. A similar logic could be applied to Siberia's hydropower resources, given the right political conditions.

There is also a more negative scenario, in which Russia instead uses heavily subsidized energy from abundant coal, natural gas or nuclear power to produce

51   Hill, Fiona and Clifford Gaddy. *The Siberian Curse: How Soviet Planners Left Russia Out in the Cold.* Washington, DC: Brookings Institution, 2003.

52   Takla, Einar. 'Nå har vi nok'. *Dagens Næringsliv*, 25 August 2008. http://www.dn.no/forsiden/borsMarked/article1475178.ece?WT.mc_id=dn_rss [accessed 29 August, 2008].

renewable energy input factors such as silicon and hydrogen. Some Russian actors are already thinking in this direction. For example, the oligarch Oleg Deripaska, who is the main shareholder of the conglomerate Basic Element and the world's biggest aluminium producer Rusal, has aired the idea of establishing polychrystaline silicon production at Sosnovy Bor outside St. Petersburg. The point would be to take advantage of the cheap energy from the nuclear plant located there.[53] Insofar as Russia's output of silicon (and, in the future, hydrogen) will continue to be oriented towards exports, the demands of the market for green energy will however give Siberian hydropower the upper hand over more centrally located nuclear power and natural gas.

The long-term investor, a rare post-Soviet species

Post-Soviet business culture was shaped by the collapse of the Soviet Union not only as a communist state but also as a moral and cultural system. Because communist ideology saturated so many levels of life, when the ideology dissolved into thin air an anarchic cultural vacuum developed. Important economic assets such as factories, oil fields, gas pipelines, stocks of valuable metals and real estate of value were up for grabs.

In this context, survival of the meanest and the best connected became the rule. Some people became immensely rich in a short timespan. As a result Moscow now has the highest concentration of USD billionaires in the world.[54] Widely known as 'oligarchs' these *nouveau riche* are accustomed to a post-Soviet culture of grabbing as many assets as they can hold, and thereafter squeezing them until the last drop. The oligarchy therefore is not oriented towards long-term investment, indeed it would like to limit investment as much as possible. This reflects both the sociological trajectory along which the oligarchy has evolved – this is simply how it is used to making money – and is further reinforced by lack of clarity about property rights both at an ethical and at a legal-institutional level. Ethically speaking, members of the oligarchy know how they obtained their assets and are therefore perennially worried that somebody else might similarly push them aside and grab their businesses. At the legal-institutional level this is reinforced by concrete cases in which oligarchs have been unable to defend their assets, most famously in the case of oligarch Mikhail Khodorkovskiy and his oil company Yukos. Many feel a large degree of uncertainty about the long-term security of their property and there is an inevitable reluctance to invest. In stead, the emphasis has

---

53   Kreutzmann, Anne. 'The Smell at the End of the World: Nitol Wants to Produce Silicon in Siberia'. *Photon International*, no. 11 (November 2007): 30–47.

54   Moscow has 87 billionaires, ahead of New York (71) and London (36). Kroll, Luisa. 'The World's Billionaires'. *Forbes Magazine*, 5 March 2008. http://www.forbes.com/lists/2008/03/05/richest-people-billionaires-billionaires08-cx_lk_0305billie_land.html [accessed 25 June 2008].

been on securing political power to protect economic assets, or on getting capital out of the country to safe havens as fast as possible, resulting in massive capital flight. Although some capital has returned in recent years, Russia still fails to draw the scale of Western investment that other emerging economies are attracting.

One of the most sensitive and long-term of all forms of investment is that in science and the development of new technology. Compared to a factory or a natural gas field, science is more difficult to fence in and put under armed guard, and the returns are uncertain and may be realized only in the distance future when it is difficult to predict whether one's contacts in the political, bureaucratic, legislative and security organs will still have any clout.

In this post-Soviet business context, oligarch Oleg Deripaska's company Norilsk Nickel is investing several hundred million USD in New Energy Projects (NIC-NEP), a joint venture with the Russian Academy of Sciences aimed at developing fuel cells, other hydrogen technology and solar power. This illustrates several things: that there is some large-scale, private investment in science for renewable energy in Russia and that hydrogen is a key area in the Russian renewables landscape. Not least, it shows the complexity of Russian business investors, since it is the same Deripaska who has proposed to make silicon with nuclear power and who is investing so much money in advanced research on hydrogen. It is not always easy to distinguish between good and bad in the Russian context, and this may be a challenge for Western actors, but also means that interesting opportunities can be hidden behind bad first impressions.

Conclusions

On the basis of the considerations made here, we believe that two forms of renewable energy merit particular attention in the Nordic countries' dealing with Russia: hydrogen and solar power. This is based on the following premises:

- Hydrogen and solar power came out top in the table of assessments of Russian and Nordic strengths and weaknesses at the beginning of this section.
- Hydrogen and solar power are both high-tech, basic-science fields, and in this book we are particularly interested in the potential specifically for *research* cooperation between the Nordic countries and Russia.
- Russia's strengths in the fields of hydrogen and solar power include research groups in Moscow and Siberia (as well as St. Petersburg), and are thus good examples of how many of Russia's renewable energy assets are located outside the northwestern part of the country that Nordic actors tend to focus narrowly on.

- The production of both hydrogen and silicon for photovoltaic applications require large amounts of clean energy. As we saw above, Siberia and its large waterways may have something special to offer in this respect.
- As shown above, both hydrogen and solar power are key sectors in the Nordic renewable energy landscape, providing a neat match with Russian interests.
- Hydrogen and solar power have attracted investments both from the Russian state and private Russian investors on a scale that is unique in the Russian context.

On the background of these considerations, hydrogen and solar power have received special attention in the course of the work on this book. A presentation on hydrogen was made at the conference *Renewable Energy in Russia* in May 2008, and is available on the internet.[55] Russia's solar power sector is discussed in Chapter 4.

What the matching of comparative advantages exercise in this chapter has shown is that actors seeking to engage in renewable energy partnership with Russian partners should not let the geographical or technological biases of their home country decide the focus of the collaborative ventures. Other factors should be considered as well. The case of the Nordic countries has revealed an understandable tendency to let own capacity influence decisions over cooperation while largely overlooking the competence of the Russian side. The exercise also reveals the benefits of our social science approach to the matter: Not only investment capacity and resource base matter. Basic science competence is also relevant. So are markets, infrastructure and other structural conditions.

55 Øverland, Indra. 'Russia's Hydrogen Sector'. Presentation at the conference *Renewable Energy in Russia: How Can Nordic and Russian Actors Work Together*, Oslo, Norway, 8 May 2008. http://english.nupi.no/content/download/4431/61693/file/Indra%20Overland%20[Read-Only].pdf [accessed 22 June 2009].

Chapter 7

Working on Renewable Energy in Russia: Ten Experiences[1]

This chapter deals with the subjective experiences of ten people who have cooperated with Russian actors on renewable energy projects. Why do they find such cooperation worth pursuing? What has been the outcome of the cooperation?

The main findings from the ten interviews are that perceived opportunities for cooperation are mostly related to Russia's natural resources, mainly hydro, wind and bioenergy. However, despite the many opportunities, the respondents have also experienced significant challenges, especially in relation to cultural differences, corruption and bureaucracy. These barriers appear to hinder the concept of Nordic–Russian cooperation from materializing into concrete action and business development.

The ten actors interviewed for this chapter are located in Norway, Sweden, Denmark, Finland and Russia; and represent companies, state organizations, an NGO, a university and a financial institution. They were subjected to in-depth, structured interviews based on the interview guide that can be found in the appendices. These ten interviews represent just over 10 per cent of the interviews on which this book is based. They stand out because they were carried out as structured interviews, because they focus on the experiences of *Nordic* actors and because they are largely oriented towards the business sector. Many other parts of this book aim to identify the positive opportunities for engaging with Russian actors in renewable energy, mainly in science, and to show how large and varied the sector in fact is in Russia. We therefore thought it important to include these ten interviews in order to explore some real-life experiences from renewable energy projects in Russia, and also to bring out some of the more negative aspects and challenges. We believe the experiences of these Nordic actors can be a source of guidance for other Western actors seeking renewable energy cooperation in Russia.

Methodology

The selection of respondents for this part of the study was not naturally given. Johannesen et al. suggest a range of selection criteria depending on the research

---

1   This chapter was written by Nina Kristine Madsen.

problem.[2] In principle it would have been interesting to carry out a quantitative survey, but this was not possible for several reasons. Firstly, very few Nordic actors are in practice involved in actual cooperation with Russia on renewable energy, so the relevant population is too small. Secondly, there exists no complete or partial overview of all the actors engaged in such activities. Thirdly, we had limited time and resources at our disposal and thus were unable to generate an exhaustive overview. Therefore we chose a qualitative approach, aiming to obtain a significant amount of information from a limited number of respondents. Regarding the number of respondents, Johannesen et al. state that there is no limit to the number of interviews that can be done in a qualitative study, but indicate the ideal number as being between 10 and 15 respondents.

Our goal was to identify key respondents rather than a very large number, and to include respondents who represented a variety of perspectives and could shed light on different aspects of renewable energy cooperation. In the literature, Johannesen et al. refer to 'purposeful sampling', which means that the researcher has chosen to make the selection strategically. The criteria for the selection through such an approach are not given in advance, and might crystallize in the course of the process of working with the study.[3] The selection of respondents was therefore carried out not in order to generate representative results, but in order to identify suitable respondents.

There are various ways of conducting a strategic selection. Initially, we considered using what Johannesen et al. refer to as 'maximum variation': selecting respondents on the basis of different distinguishing marks. For instance, what country or which type of renewable resource they focus on, or which type of approach they have (company, research, NGO, political). However, the maximum variation approach requires that interviews are carried out and analysed and new respondents are chosen on this basis. With the time limit involved in this study, it was not possible to work in such a manner. As there existed no overview of whom to contact, we chose the snowball method, similar to the co-nomination process.[4] Johannesen et al. explain this as a method where the researcher tries to identify persons with a high degree of knowledge on a specific topic, and then asks them about other potential respondents of relevance. However, we did to some extent determine some criteria in advance:

- involvement in Nordic–Russian cooperation on renewable energy at some level, preferably business cooperation

---

2   Johannesen, Asbjørn, Line Kristoffersen and Per Arne Tufte. *Forskningsmetode for økonomiskadministrative fag*. Oslo: Abstrakt forlag, 2004.

3   Johannesen, Kristoffersen and Tufte 2004.

4   Dannemand Andersen, Per and Birte Holst Jørgensen. *Grundnotat om metoder indenfor teknologisk fremsyn.* Risø: Forskningscenter Risø, 2001, p. 7.

- variation in location (Norway, Finland, Denmark, Sweden, Iceland, Russia)
- variation in the renewable energy source in focus (wind, hydro, bioenergy).

When we started the search for respondents, we expected to find quite a number of companies with this type of experience. However, we soon realized that this was a relatively new field for business cooperation. There seemed to be an evolving interest among companies, with an increase in conferences both in the Nordic countries and in Russia, but not much practical experience. We therefore decided to include other organizations with considerable experience of Nordic–Russian cooperation in the energy field, as these were most likely to have broad knowledge about opportunities and challenges for cooperation on renewables.

We ended up with ten respondents located in Norway, Russia, Sweden, Finland and Denmark. They have all been involved in various types of Nordic–Russian cooperation, mainly within the field of renewable energy, although one of the governmental organizations was involved in energy cooperation with Russia on a more general level. This type of small-N, qualitative survey is best suited for exploratory purposes, which is also our aim here.

In spite of the qualitative nature of the ten interviews, we have chosen to present some of the material in tables that can offer a more quantitative overview of the interviewees' answers. This does not mean that the material is intended to have quantitative validity, but rather that we hope this sort of quick overview of responses to a specific question can be of use to those readers who are particularly interested. The ten interviewees may be a small group, but they are some of the most experienced people in Nordic–Russian cooperation on renewable energy.

Perceptions of Russian business culture and environment

Although we will be contrasting Russian and Nordic business cultures, it must be made clear that Russian business culture is neither monolithic nor static. There are many different actors with different approaches to business in Russia, and the situation is constantly changing. And of course the Nordic countries do not have a monolithic or static business culture either, but many different, varying business cultures. That said, we still think there are a range of systematic differences or tendencies that set most Russian and Nordic businesses apart. Perception can play a considerable role, for there are areas where the disparity between business cultures is rather small, yet differences are perceived to be great. For most Western small and medium-sized enterprises, the Russian business environment is still unknown territory, nearly two decades after the fall of the Soviet Union. Russia does not have the same democratic tradition as the Nordic countries, and is characterized by a history of authoritarian and totalitarian rule. This is perceived to have influenced Russian business culture, just as strong democratic traditions have influenced

Table 7.1   Perceived differences in business culture[5]

Nordic countries	Russia
Informative	Communicative
Product-oriented	Relational
Planning	Intuition
Neutral	Personal, emotional
Democratic	Hierarchical
Honest	Seize the opportunity
Non-corrupt	Get around systemic obstacles

the business culture of the Nordic countries. Regarding the business culture, differences are perceived to be related to predictability, bureaucracy, legislation and contradictory rules. Russian laws can be extremely detailed, and may or may not work in practice.[6]

According to Hjelm,[7] the differences between Russian and Nordic business cultures are especially apparent in decision-making processes, during the process of business development, and in client–supplier relations. An overview of some of the major differences is provided in Table 7.1.

These potentially significant differences in business culture indicate the importance of cultural knowledge and understanding.[8] Such differences are also emphasized by Swahn (2002), who concludes that the most common differences between Nordic and Russian cultures lie in the hierarchic structures of society and power distribution in the latter.[9] Regarding planning, Swahn notes that in Norway it is essential to follow the schedule and it is common to be result-oriented, while the Russian way of planning is more short-term and contextually conditioned.

Business cultures may also differ regarding the role of leaders. While a good Nordic boss is seen as someone who is democratic and a team facilitator, a good Russian boss is assertive and capable of making swift decisions. Additionally,

---

5   Source of data used to compile table: Hjelm, Henrik 'Russian Business Culture: To Do Business in Russia'. Presentation at the conference *Investing and Financing Renewable Energy in Russia*, Stockholm, Sweden, 15–16 April 2008.

6   Bond, Derek and Markku Tykkylainen. 'Northwestern Russia: A Case Study in "Pocket" Development'. *European Business Review*, vol. 96, no. 5 (1996): 54–60.

7   Hjelm 2008.

8   Hjelm 2008.

9   Swahn, Natalia. *The Role of Cultural Differences between Norway and Russia in Business Relationships: Application to Strategic Management in Norwegian Companies*. Ph.D. dissertation, Norwegian University of Science and Technology, Trondheim, Faculty of Social Sciences and Technology Management, 2002.

personal relations are accorded even higher priority in Russia than in the Nordic countries (although they are also salient there), and it is crucial to establish such relations in order to come to an agreement on a project or contract.[10] Awareness concerning these differences is vital when working together with Russian actors. How then are the possible synergies between Nordic and Russian actors and the great potential realized? The next section presents experiences from Nordic–Russian cooperation on renewable energy.

Respondents

In the following section we describe the ten respondents, their role and relation to Nordic–Russian cooperation on renewable energy, the organizations they represent and the positions they hold. The respondents are not mentioned by name.

The Nordic Council of Ministers opened a new office in Kaliningrad in 2006. A Nordic Director and several local personnel run the office. The aim of the office is to identify relevant Russian partners, facilitate Nordic–Russian cooperation and establish networks with the national authorities, NGOs and other national and international players in the Russian regions. Our respondent is the director of the Kaliningrad office and has held this position since its opening. This person has also been involved in energy-related cooperative projects, in addition to having worked on projects elsewhere in the Baltic region and the European part of Russia since 1995.

The Swedish Energy Agency was established in 1998 with the aim of transforming the ecological and economic sustainability of the country's energy system. Additionally, the agency guides state capital within the field of energy, through cooperation with industry, energy companies, municipalities, research institutions and through trade. Internationally, the agency is active in various forums and has been involved in cooperation in the Baltic states as well as in Russia. Our respondent works at the Swedish Energy Agency, currently with the department of administration and finance, but has been involved in many energy projects in Russia and the Baltic states. Approximately 45 of these projects involved installation of biofuel boilers to replace oil-fuelled ones. From 1997 to 1998 the respondent was involved in cooperation in the Baltic region on the industrial level, dealing with the implementation of Nordic products in the energy sector in Russia. This person later headed a Nordic group within BASREC, working on bioenergy cooperation in the Baltic region including Russia.

The Nordic Council of Ministers, with headquarters in Copenhagen, was formed in 1971, and acts as a forum for Nordic governmental cooperation. The

---

10   Swahn 2002.

Nordic countries work together in a range of areas and with many countries, including energy and Russia. Two respondents were interviewed, one with expert competence on energy cooperation, the other one with expertise on collaboration with Russia.

Norwegian Wind Energy is a consultancy and project company which aims to establish renewable energy production at suitable sites in Norway and abroad. The company has more than 10 years of experience in working with issues concerning wind power, and is also the biggest shareholder in the Norwegian–Russian joint venture VetroEnergo. Our respondent has followed developments from the very beginning. Additionally, this person has a decade of experience from cooperation with Russia through the establishment of windmills in Murmansk, and was actively involved in the establishment of VetroEnergo.

Statkraft is a state-owned Norwegian company that is a leader in Europe in the field of renewable energy. The company generates hydropower, wind power and district heating, and has constructed gas power plants in Germany and Norway. Additionally, Statkraft is a major player in European energy exchanges. Regarding Russia, the company has a range of ongoing collaboration projects on hydropower, and has had a Russian partner since 2002. Its main Russian partner is Hydro OGK, with which it has a cooperative platform. Our respondent heads the company's Russian activities, and has been responsible for the calculations and financial viability of Russian projects.

Tricorona is a Swedish-owned company with a subsidiary in Moscow. The company has been listed on the Stockholm stock exchange since 1989, and is the second largest buyer of climate development mechanisms worldwide. Its main business today is investments in carbon credits, with activities carried out within the framework of the Kyoto Protocol. Our respondent started a Swedish–Russian joint venture within this field three years ago, but realized that it was too small to operate alone. The original company was then sold to Tricorona in 2006. The respondent is responsible for Eastern Europe, is fluent in Russian and is currently working in Russia.

Rosnor is a Norwegian–Russian joint venture working with Russian power companies with plans to develop hydropower plants and other renewable energy projects. Additionally Rosnor is working on a project with the Norwegian Ministry of Petroleum and Energy aimed at establishing cooperation between the two countries in order to promote small hydro investments in Russia. Our respondent has long experience from the Norwegian and Russian hydropower sectors, works partly in Moscow, and speaks Russian.

The Nordic Environment Finance Corporation (NEFCO) is an international financial institution established in 1990 by Finland, Denmark, Norway, Sweden and Iceland. To date, NEFCO has financed a wide range of environmental projects in Central and Eastern European countries, including Russia, Belarus and Ukraine. NEFCO's headquarters are in Helsinki and its activities are focused on projects that achieve cost-effective environmental benefits across the region. Our respondent has been working in NEFCO since 2002 as a Special Advisor on Energy and the Environment. The respondent has experience from the financial sector, and has previously worked for the World Bank's office in Moscow, involved in their establishment of the project 'Russia's renewable energy programme'.

Bellona was formed in 1986 as an environmental, non-profit foundation. It has primarily worked on nuclear contamination in Russia, but has also carried out studies on the potential for renewable energy on the Kola Peninsula. In the spring of 2008 it launched a new initiative for Nordic–Russian cooperation in the field of renewable energy. Our Bellona respondent is Energy Advisor at the foundation's Russian Department and is currently working on the establishment of a Nordic–Russian Forum for renewable energy as a platform for further Nordic–Russian cooperation within this field. The respondent has broad knowledge of Russia and speaks Russian.

The Swedish University of Agricultural Science (SLU) aims to develop understanding and sustainable use of biological and natural resources. Its profile areas are genetic resources and biotechnology, climate and ecosystem change and sustainable production. SLU has engaged in bioenergy cooperation with Russia since 2000. Our respondent works at SLU as a Professor in Bioenergy and has been involved in bioenergy projects and cooperation with Russia for the past eight years.

Further details about the respondents are presented in the appendices.

Opportunities for utilizing renewable energy in Russia

When we asked about the opportunities for utilization of renewable energy sources, the interviewees spoke about the whole range of renewable energy sources, including hydro, wind, bioenergy, tidal and solar. Many viewed hydro as the source with the greatest potential in Russia, but wind and bioenergy were also emphasized. Bioenergy was mentioned in relation to the enormous forestry resources in Russia, and some respondents opined that there could soon be commercially viable projects. The opportunities mentioned were often related to Northwest Russia, where many of the respondents were operating. Statements like the following were rather common:

> Russia is a big country. I will focus on Northwest Russia. All hydro could be possible; when it comes to wind there might be possibilities in Murmansk and Kaliningrad.

Table 7.2　Which types of renewable energy are most promising in Russia?

	C1	C2	C3	C4	A1	A2	A3	F	U	NGO	Sum
Hydro	•	•	•	•	•	•		•		•	7
Wind	•	•	•		•	•		•		•	6
Bioenergy						•		•	•		3
Tidal		•								•	2
Solar										•	1
No reply							•				1

The interviews revealed that all the companies thought first and foremost of hydropower in connection with opportunities in the renewable energy sector in Russia; by contrast, the university, the financial institution and one of the state organizations focused on bioenergy.

Several respondents noted that there could be seasonal complementarities between wind and hydro. This was reasoned from the fact that there is snow rather than rain during wintertime, but good winds at that time of the year. Conversely, summertime is characterized by little wind but good possibilities for rainfall. The responses are shown in Table 7.2.

Why is cooperation interesting?

As to why Nordic–Russian cooperation in the field of renewable energy is of interest, an essential feature appears to be the *potential* of renewable energy in Russia as well as the business opportunities these resources represent. Answers like this were not unusual: 'You have to go where the potential is, and certainly in Russia there is a huge potential.'

The great potential of resources seems to serve as a foundation for cooperation opportunities. This opinion was evident from some of the other respondents as well, but did not appear to be obvious to all. Atypically, one respondent said that Nordic–Russian cooperation on renewable energy *is not commercially interesting* for Nordic actors. The respondent's reasoning was as follows: Russia may import technology, for instance, much more cheaply from China. This respondent argued that the Nordic countries do not have a great competitive advantage. Differences of opinion among the interviewees on this question indicate the varied nature of Nordic–Russian renewable energy cooperation.

Some respondents emphasized the role of Russia as a neighbouring country, and one put it this way:

Table 7.3   Why do you think Nordic–Russian cooperation is interesting?

	C1	C2	C3	C4	A1	A2	A3	F	U	NGO	Sum
Potential of natural resources	●	●	●		●	●				●	6
Political					●				●		2
Neighbours	●		●	●							3
Profit			●		●			●	●		4
Exchange of knowledge							●		●		2
Not commercially interesting				●							1

> The Nordic area is the only area in the western world that borders directly on Russia. We are linked to Russia, whether we want it or not. It's a neighbour. I think we should turn the question around and ask: why we should not have a business relationship with a neighbour? That would be more surprising.

This statement shows that the interest also may be founded on geographical proximity, where companies might save themselves the need to travel halfway around the globe for new markets. Some respondents added that they were surprised by what they characterized as low Nordic interest in collaboration with Russia on renewable energy. This can possibly be explained by structures for financing. A key factor for the low interest may be that the Nordic governments have generally accorded priority to renewable energy cooperation with countries elsewhere in the world, as in Chile, China, Brazil and Africa. This may indicate that Russia falls between Nordic priorities.

Furthermore, the responses from the university and one of the state organizations indicated *exchange of knowledge* as one reason for their interest. Here it should be noted that it was *exchange*, not *transfer*, of knowledge that was mentioned. These respondents spoke of the opportunity to learn from the Russians, as well as teaching them. The responses are shown in Table 7.3.

Initiating the cooperation

As for the *initiatives* for the cooperation, these can be summed up in one word: *mutuality*. There was a good mix between initiatives coming from the Nordic and the Russian sides. This is perhaps surprising and also interesting, as our original expectation had been that the Nordic side would prove to be more active than the Russian side.

The respondents have in many cases searched actively, and some could tell quite similar stories about meeting Russians who were as keen to initiate similar cooperation as they were. These trends may serve as a good basis for further cooperation within this field.

Table 7.4   In what areas do you think there are opportunities for cooperation?

	C1	C2	C3	C4	A1	A2	A3	U	FI	NGO	Sum
Hydro	•	•	•	•	•			•			6
Wind	•	•			•			•			4
Bioenergy					•			•			2
Russian projects/production			•		•						2
Tidal		•									1
Transfer of technology			•								1
Export of Nordic knowledge										•	1
Northwest Russia						•					1
Construction										•	1
Subcontractors					•						1
All sorts of possibilities									•		1
Don't know								•			1

Opportunities for Nordic–Russian cooperation

As we have seen, the interviewees indicated hydro, wind and bioenergy as areas of potential interest in Russia. The tendencies appear to be independent of which of the groups the respondents represent. One respondent put it this way: 'By looking at the bare figures, you can see that for companies focusing on environmentally friendly energy sources, Russia is clearly a target. It's as simple as that.'

In general, respondents tended to be very optimistic as to the opportunities for Nordic–Russian cooperation in the field of renewable energy. Several specifically mentioned the great opportunities based on the Nordic actors' knowledge and experience from the various renewable energy sectors. A few respondents also noted the opportunities for financially profitable projects in a short-term perspective, and one asked this rhetorical question: 'Why not try to conquer a market that is practically virgin territory in terms of certain types of technology?'

To summarize, the interviews indicate that there appear to be good opportunities for Nordic–Russian cooperation within several fields, especially hydro, wind and bioenergy. But even though the respondents pointed out many different opportunities, many quickly moved from talking about opportunities to the challenges and barriers involved. On the basis of the interviews, we are convinced that there is a high awareness of the opportunities among the respondents, and these seem directly linked to the potential of the various resources. At the same time there is also a high awareness of the barriers to be overcome before one can take full advantage of these opportunities. Table 7.4 summarizes this section.

Motivation/goal

Motives for Nordic–Russian cooperation on renewable energy varied. The companies and one of the state organizations stated that their goal was to achieve concrete results through the projects. From a company perspective, this is only to be expected, as their activities must be commercially viable. Interestingly, however, one of the respondents from the governmental organizations also noted this as a goal. While many respondents have quite clearly defined goals, this does not seem to be the case for all of them, and one respondent put it this way: 'We don't have a defined goal but we want to see as many concrete results as possible.'

The NGO and one of the governmental organizations said their motives/goals were ideological, and this was exemplified with the motivation for reducing global warming. The two other representatives of the authorities replied that their motivation for cooperation related to political motives, exemplified with energy security goals and involving Russia in a dialogue. The university and one of the companies emphasized knowledge transfer, while the financial institution simply tries to stick to the budget. The overall impression of Nordic–Russian cooperation on renewable energy is that motives/goals differ among the respondents, and the motivation and goals reflect the organizational context in which they operate.

Time horizon

As for the time horizon, practically all interviewees stated that the time horizon for their involvement in Nordic–Russian cooperation on renewable energy was long-term. As one respondent noted: 'We have been working on this project for 10 years, and it has been more like a walk in the desert. Wasn't Moses in Sinai for 40 years?'

None of the other respondents had been involved for so long, but almost everyone underlined the need to be very patient. However, it was also pointed out that this was usually not a problem for actors working on renewable energy, as many of them do not see their sector as a fast track to profits anyway.

In the following sections we account for the process of organizational cooperation, including partner selection, Nordic–Russian complementarities and sources of guidance and norms. Furthermore, we examine the challenges and barriers to Nordic–Russian cooperation on renewable energy.

Partner selection

As for partner selection and the companies, one got an invitation to cooperate, while the other three companies referred to their partner selection as a mix of *coincidence* and *choice*. Several respondents underlined the role of their Russian cooperative partners as 'interpreters' of the Russian system. In general, the Russian partners seem to be of great importance to the Nordic actors. *Trust* was also discussed under this topic, but seems relevant mainly to the companies. Our reflection on this is that it is mainly the companies and the financial institution that are involved in significant economic risk, while the others respondents function more as initiators or facilitators. A respondent from one of the companies put it this way: 'The hardest thing in a new country is to find people you can rely on.'

Some other companies mentioned the importance of trust between partners. However, it was also emphasized that this is nothing unique to Nordic–Russian cooperation on renewable energy, but is a feature common to all kinds of business relations. As noted by one respondent:

> Whenever we sit down together in order to draft the legal documents you know you have to be aware of the opponent. If you don't do your homework you might be trapped, but this does not have to do with trust, it's business.

Another respondent was clear about the fact that they were actively involved in trying to bring Nordic and Russian companies together in a form of 'matchmaking'. When asked a bit more about this topic, the respondent explained that they organized workshops and conferences in relation to renewable energy where Nordic and Russian actors could meet. The university, the NGO, the financial institution, one of the governmental organizations as well as one company had all contributed to partner selection processes.

The general impression from the interviews is that the respondents in this study have been able to establish good relations with their Russian partners, with the selection generally characterized by a mix of active search, coincidence, choice and invitations.

Nordic–Russian complementarities

Regarding complementarities between Nordic and Russian actors within the field of renewable energy, the factors underlined were the local knowledge and potential resources on the Russian side, and the competence and technical know-how in the Nordic countries. It is noteworthy that company 2 (C2), along with the university, stressed knowledge exchange, while the NGO underlined its ability to bring in funding. Normally one would expect that companies would focus more on funding, and the NGO more on knowledge. This finding is interesting and may help create new opportunities for Nordic–Russian cooperation, as the respondents seem to find complementarities within other areas than the apparently 'natural' fields. The responses are shown in Table 7.5.

Table 7.5   How do you and your Russian partner complement each other?

Respondent	
C1	Interested in cooperation on renewable energy at a very early stage (1998), the Russian partners were looking for interested partners
C2	Knowledge exchange
C3	Local Russian knowledge and political contacts vs local Norwegian contacts and knowledge
C4	In no particular way
A1	Lack of technology and knowledge in Russia. Nordic countries have a lot to offer
A2	Enormous resources in Russia, technical knowledge and equipment in the Nordic countries
A3	Nordic countries have a lot of competence on renewable energy and the Russians seem interested in cooperation
U	We had technical and economic knowledge on forestry issues, the Russians had a narrower competence, but knew other things about different technologies
FI	You need a local partner in Russia
NGO	They have the local know-how, we are bringing in the funding

Sources of guidance and norms

We sought to learn more about what institutions the respondents regarded as important sources of guidance and norms. The national institutions of each of the Nordic countries were one important source. Examples of such organizations are the Danish Trade Council, Innovation Norway or the Swedish Energy Agency, as well as the national embassies. Many respondents have good relations with one of these, and have received help at some level. In general, the respondents seemed satisfied with the help and guidance provided, but the various national organizations appear to have little coordination and information sharing in this field. This might be a challenge, as it could indicate lack of coordination and knowledge-sharing among national Nordic institutions. One respondent asked for more cooperation among the national institutions of the Nordic countries. A heightened business focus from the embassies was also desired.

Additionally, the interviews revealed that Russian institutions, both in Moscow and at local/regional level, are seen as important. This was also the case with the Russian partner. One respondent made this reflection:

> In the Western part of Europe we have institutionalized our relations. In Russia it's more person-to-person, and if you manage to establish this kind of relation, that is how you do business. When you get such a relation it is strong but vulnerable, because if someone quits or is replaced, it's back to square one.

Table 7.6   What institutions provide guidance for the cooperation?

	C1	C2	C3	C4	A1	A2	A3	U	FI	NGO	Sum
National Nordic institutions			●	●		●	●				4
Nordic financial institutions		●	●	●							3
Local/regional authorities	●								●	●	3
Russian partner		●							●		2
Central institutions in Moscow	●								●		2
Universities								●			1
Russian energy associations						●					1
Sector regulators		●									1
BASREC									●		1
NGOs									●		1

The Russian person-to-person mindset was mentioned by many respondents. Several interviewees made statements to the effect that that there is no chance of survival without a partner you can trust, and a partner that can interpret the Russian context for the Nordic actor. The responses are shown in Table 7.6.

Challenges and barriers

No matter how bright the prospects may appear, there are some barriers to overcome in Nordic–Russian cooperation on renewable energy. Our respondents seemed to have a shared understanding of this, regardless of which of the Nordic countries they represented, and whether the replies came from companies, the non-profit environmental organization, the financial institution, the university or governmental organizations. We have divided the barriers into two categories. The first category is related to the Russian context as the Nordic actors experience it, while the other category is related to the barriers the Nordic actors experience within the Nordic countries. The interviews showed that many barriers seemed to relate to the Russian context, but there were also diverging opinions.

In fact, barriers were a topic that often came up when discussing the other areas as well. The overall impression is that there seem to be a significant number of barriers. Among the respondents, eight out of ten stressed culture as a barrier to Nordic–Russian cooperation on renewable energy. This included all the companies, the university, the NGO, one of the governmental organizations and the financial institution.

Another barrier related to the Russian context was corruption. This was mentioned by six respondents, and the interviews left little doubt that this is seen as a challenge in terms of cooperation. The respondents confirm that they see Russian business culture as rather different from the Nordic way of doing business. Some said that corruption had never been a problem, but added that they used various kinds of 'middlemen' in order to deal with permits, applications and help them to manoeuvre around in the Russian system.

Two respondents stated clearly that Nordic actors lack knowledge and competence on the Russian market and that this is a great barrier. They also made it clear that if Nordic actors within Nordic–Russian cooperation view culture and corruption as barriers, no progress will be made, because it is difficult or impossible for one company to change the entire Russian system.

Four out of ten respondents mentioned Russian governmental bureaucracy as a barrier to Nordic–Russian cooperation on renewable energy. Many said that this was related to getting licenses and permits, but mention was also made of the rapid turnover of officials within the bureaucracy, as the person-to-person relation is so central. One respondent reflected on the challenge with the bureaucracy in this way:

> Even if there is a market, even if there is a demand, if you are a foreign operator in the country and someone is putting stones in your wheels all the time, you might realize some day that there is a price for this market and that it's not competitive. The cost of this can be too much. Then you won't do business there.

It seems logical that culture, corruption and bureaucracy get similarly high scores as barriers, as these are areas that are closely related to each other. One respondent emphasized the following, which can be an explanatory factor as to why some respondents experience the barriers mentioned above:

> Nordic companies have spent too little time trying to analyse and understand Russian culture. The knowledge of Russia itself is missing. One barrier is lack of knowledge about Russian society, Russian financial systems, Russian economics, politics, everything. Knowledge about Russia is in general poor.

Cultural barriers are often mentioned as barriers to many types of business cooperation, and might sound like a cliché. However, the responses leave little doubt as to its importance in Nordic–Russian cooperation on renewables.

Another cultural factor which some respondents mentioned as a barrier is language. However, the overall impression from the interviews was that communication between Nordic and Russian partners was solved reasonably well by using translators and interpreters, communicating in English or by having Russian-speaking staff within their organization. Regarding language, one respondent made an important point:

> If you go to Denmark and do not speak Danish or English you have a big
> problem. Same thing with Russia. There are a lot of business areas in Russia
> where they do not speak English, but it is exactly the same if you go to China,
> Japan or Congo. This is not Russia-specific. If you don't speak the language it's
> impossible. Try that in Congo. It's common sense.

Seen from this perspective, language is a barrier that can be perceived as being
either a problem on the Russian side (that they do not speak English) or a problem
on the Nordic side (that they do not speak Russian).

The time aspect was also mentioned as a barrier. The *slow progress*, meaning
the time it takes to develop projects, is cited as a barrier to Nordic–Russian
cooperation on renewable energy. Respondents were quite frank, telling stories
about how incredibly long it might take to develop projects, like the one who gave
this response:

> I don't know if we would have started if we had known it would take such a long
> time. However, you are always optimistic and look for signs that will tell you
> that it will change. Now we are looking for the 12th sign, hoping that something
> sustainable will occur.

The slow progress is related to bureaucratic obstacles, and to the general Russian
view of time. One respondent noted the Russian tendency to view time as something
that is free, while the Nordic actors hold to the adage that time is money.

Regarding other barriers that the respondents have experienced, there are also
significant challenges related to legislation in the field of renewable energy. Many
noted the need for a more comprehensive framework. Laws often seem to be more
like strategic documents than something concrete, especially when it comes to
tariffs for support schemes and related matters. A few interviewees mentioned
this area of barriers in Nordic–Russian cooperation, but it was also noted that the
year 2008 could be a milestone, as Russian policy-makers are about to make new
provisions for concrete support schemes for renewable energy.

Also similar to the barriers in legislation, the respondents cited concerns
regarding access to information and people. It can be difficult to get access to
statistics, information and people in state organizations in Russia. Several
respondents mentioned the need for a publicly available overview made in one or
more of the Nordic countries. This can also, however, be interpreted as a transparency
issue. While Nordic partners are used to publicly available information, this is less
common in Russia – and it is easier for Nordic actors to call for an overview of
Russian public information to be made by Nordic coordinating agencies than for
the Russian system to change in a more transparent direction.

Lack of predictability is another area noted as a barrier in the Russian context. Only two respondents mentioned this explicitly, and both were from companies with commercial ambitions. One respondent explained:

> We can deal with the market risk, the price risk, the volume risk and whatever business risk, but whenever the regulatory risk changes rapidly and you can't anticipate what to expect around the next corner, it is tough to undertake the kind of investment that is needed.

Problems with the electricity grid appear to be a more notable barrier, as the lack of a strong grid in Russia influences the possibility of transporting the generated electricity from the north and further south to areas with high demand. This is perhaps a structural barrier in the Russian context. However, it can also be interpreted as a failure on behalf of the Nordic actors to adequately foresee such problems before starting projects. As will be discussed in the conclusion of this book, we found the Nordic focus on setting up generating facilities not to be the best way of realizing the potential that exists within Nordic–Russian renewable energy cooperation. Russian technical standards are also a field that two respondents mentioned as a barrier. Russia, unlike the countries in the Baltic area, has its own standards for technical equipment and does not follow EU standards. This may complicate things, and create delays. As one of the respondents put it: 'It has always been much easier to cooperate with the Baltic countries, even long before they became EU members. Russia is much more difficult.' On the other hand, there is a considerable difference between the market potential of Russia and the Baltic countries, due to the sheer size of the former.

Despite the various barriers to Nordic–Russian cooperation on renewable energy, it seems appropriate to end this section by quoting one of the respondents and summarizing the challenges and barriers as set out in Table 7.7. 'Get away from the barrier-thinking and move on to opportunity-thinking. That's where the opportunities are.'

Main challenge

When we asked respondents to rank the challenges in Nordic–Russian cooperation on renewable energy, we got seven different answers, highlighting very different aspects. Seen in the relation to the question about barriers, one might have expected culture to be stressed here as well. However, that was not the case. Only one respondent considered this to be the main challenge. In the responses to the previous question about the barriers, no one mentioned the power of Moscow as a barrier to Nordic–Russian cooperation, even though, from the answers presented below to the second question about barriers, this emerges as the main challenge for three of the ten respondents. The differences in the answers could possibly be due

Table 7.7   What does not work in existing Nordic–Russian cooperation?

		C1	C2	C3	C4	A1	A2	A3	U	FI	NGO	Sum
RUSSIAN	Culture	●	●	●	●		●		●	●	●	8
	Corruption	●	●	●			●	●		●		6
	Bureaucracy	●					●	●			●	4
	Slow progress		●		●	●	●					3
	Lack of predictability	●	●									2
	Language		●				●					2
	Problems with the grid	●									●	2
	Access to people and information	●									●	2
	Russian technical standards						●			●		2
	Low Russian demand and awareness for renewable energy					●				●		2
	Poor Russian investment climate									●		1
NORDIC	Nordic actors lack knowledge and competence on Russian culture and society	●	●								●	3
	Cooperation not systematic enough						●					1
	Nordic equipment companies too small							●				1
	Lack of political focus											0

to the way the questions were posed, and that this section invited respondents to describe the major challenge in more general terms. Possibly, the power of Moscow can be seen in relation to the power structures within the Russian energy sector, as previously described in the context chapter. It can also simply be the sum of all the opaqueness prevailing in Russia, which, to Nordic actors lacking experience of working in Russia, can be perceived as coming from above – something which is not necessarily the case.

A few respondents stated that the market price of electricity in Russia played an important role, and two noted the lack of cultural knowledge in the Nordic countries as the main challenge. The perceived main challenge for Nordic–Russian cooperation in the field of renewable energy is summarized in the list below. Respondents were allowed to state only one main challenge, and the challenges in the list are sorted according to the number of respondents that gave that answer.

1. power of Moscow
2. consumer willingness to pay

3. lack of Russian political will
4. culture
5. lack of transparency
6. lack of cultural knowledge among Nordic actors
7. realization of the potential.

Certainly, the interviews only show us what is the perceived main challenge on the Nordic side of the cooperation. But, viewed together with what we know from the context chapter about the Russian electricity sector, it becomes clear that the main challenges with regard to Russian renewable energy development remain political. Remedying them will require a change in political priorities and a clear political focus and strategy for the development of renewable energy. Nevertheless, for Nordic actors with already existing cooperation projects, cultural barriers are perceived as important.

Suggestions for improvements

The answers to our question about factors that could facilitate further or improved Nordic–Russian cooperation on renewable energy were divergent. The various recommendations for improvement revealed several suggestions for areas that could be improved in order to generate more concrete results. Several respondents mentioned funding as an area in need of improvement to facilitate cooperation. Some wanted a renewable energy fund in order to stimulate activities and research as well as greater possibilities for state-supported funding for Nordic–Russian projects in renewable energy. The second area where there are suggestions for improvement is that of access to information on Russia. Three respondents emphasized the need for information and wanted this to be publicly provided in the Nordic countries. For instance, one mentioned the importance of having an overview of the laws regulating the various renewable energy sectors, in addition to an overview of comparative regulators and public offices in Russia. This is important, as companies in all the Nordic countries could benefit from such information. Moreover, such publicly provided information could be helpful in overcoming various barriers. The two other areas mentioned by more than one respondent concerned legislation, exchange of experiences and more synergies between the Nordic countries. The main tendencies are shown in Table 7.8.

Conclusions

The main findings from the ten interviews may be summarized as follows: Regarding *concepts*, renewable energy seems to mean different things to different respondents. Nordic–Russian cooperation is of particular interest due to the great

Table 7.8   What could facilitate Nordic–Russian cooperation?

	C1	C2	C3	C4	A1	A2	A3	U	FI	NGO	Sum
Funding	•			•	•					•	4
Legislation	•									•	2
Better support systems for renewable energy in Russia		•									1
Synergy effects and exchange of experience between Nordic actors.					•	•					2
Long-term strategy									•		1
Increased Nordic willingness to invest in renewable energy				•					•		2
Public access to updated information and regulatory offices		•	•	•							3
More visible commitment by the Russians							•				1
Don't know								•			1

potential of resources in Russia and the business opportunities this creates. Most cooperation is initiated mutually by the Nordic and Russian partners.

Partner selection is dominated by choice and coincidence. We have noted possible complementarities between Russian local know-how and natural resources on the one hand and Nordic market and technical competence on the other hand. Nevertheless, the process of cooperation faces a range of barriers.

As to the structure of Nordic–Russian cooperation on renewable energy, our respondents are involved in many different forms of cooperation. Better coordination is desired among Nordic actors, and there seems to be general optimism regarding the future.

We have seen that the main opportunities for Nordic–Russian cooperation on renewable energy are thought to relate to the sizeable potential of renewable resources in Russia, especially in the field of hydro, wind and bioenergy. The *concept* of Nordic–Russian cooperation on renewable energy appears well institutionalized and also well acknowledged among the respondents, who even take the concept for granted. Nevertheless, this is not the case for the field as a whole. From the results of this study, the *process* of Nordic–Russian cooperation on renewable energy seems motivated by business opportunities and appears to involve a balance between Nordic and Russian partners regarding partner selection and initiatives.

The respondents face various barriers which seem to hinder the concept of Nordic–Russian cooperation from materializing into concrete business results and commercial activities. The main barriers identified in this study relate to the context in which the cooperation takes place. The power of Moscow, culture, corruption and bureaucratic barriers emerged as the main barriers. While these four areas might sound like old news, this confirms the view put forth by Luo,[11] that cultural distance within international cooperation between organizations is of significant importance. It may not come as a surprise, but the results of this study show that culture cannot be ignored when participating in Nordic–Russian cooperation on renewable energy, and that knowledge of Russian language, economy and society is crucial to succeed. This finding is of paramount to any international actor engaged in renewable energy cooperation in Russia, and it is our hope that this chapter can make the Nordic experience a learning experience for others.

Our ten interviewees have had mixed experiences in seeking to engage with Russia on renewable energy. Indeed, they remain undecided as to whether it is worth the effort. On the other hand, if you want to make it big in a market, you need to be there before everyone else realizes that it is attractive. Should the legal and institutional framework in Russia improve, some Nordic actors are going to be very well placed to take advantage of this.

---

11   Luo, Yadong. 'Building Trust in Cross-cultural Collaborations: Toward a Contingency Perspective'. *Journal of Management*, vol. 28, no. 5 (2002): 669–94.

Chapter 8
Conclusions

This book has aimed to give an overview of the Russian renewable energy landscape that can help international actors in formulating policies to promote international cooperation with Russia on renewable energy. We have given an overview of Russian research institutions, sectors and locations that might be worth targeting for collaborative ventures on renewable energy. Both the strengths that Russia has in the renewable energy sector and the obstacles to renewable energy research and investment in the country have been highlighted. Our conclusions and policy recommendations are based on our belief that cooperation should be about mutual benefit in science and innovation through complementarities. This can be either through the joint development of scientific knowledge, or through complementarities between scientific knowledge, natural resources, capital, management skills and/or markets.

It is widely assumed that, since Russia is no longer poor and weak but slightly wealthier and significantly more assertive, joint activities must also move from aid modus to joint financing. A more sensible approach would be a gradual transition from charity-oriented aid to aid aimed at mutual benefit, with joint financing as an ambition rather than a requirement. The EU recommends that cooperation with countries such as Russia take place at the multilateral level in order to avoid duplication.[1]

This concluding chapter ends by drawing up three scenarios for renewable energy development in Russia, to give international actors interested in cooperating with Russian actors on renewable energy some ideas about possible future trajectories of Russia's renewable energy sector, including both optimistic and more sober possibilities. A 'wild card' is also thrown in to illustrate how an unlikely event which cannot be predicted might have the potential to drastically change Russia's role in the global renewable energy sector.

Basis for international cooperation

In Russia, commercial, sociological and political approaches to renewable energy have had low priority. This strengthens the relative importance of fields where basic science plays an important role, for example hydrogen or solar

---

1   EC. *The European Research Area: New Perspectives*, Green Paper. Brussels: ECf, 2007b, p. 20.

power. For international cooperation, the complementarities become obvious: strong Russian basic research in the natural sciences, foreign skills in social science, commercialization and marketing. In our mapping of Nordic–Russian complementarities, hydrogen and solar power – both of which are high-tech, basic-science fields – emerge as the best match of Russian and Nordic strengths in renewable energy. This may also be applicable to other countries.

Russia has a vast potential within renewable energy due to its size and geographical diversity, but this potential is hardly utilized at all. It is only within large hydro, which is often not counted as a renewable energy source, that Russia has utilized a significant share of the existing potential, although also here there is much unused capacity. Therefore there are many potential linkages, and Russia has the natural preconditions to develop many different forms of renewable energy. Countries around the world should be able to find something of interest to them in Russia.

Why renewable energy in Russia?

We have argued that there are several reasons why Russia, despite its natural abundance in and great focus on hydrocarbons, coal and nuclear power, should pay more attention to improving energy efficiency and developing renewable energy. First of all, it is clear that Russia can benefit economically from according greater priority to renewable energy sources, since this will increase its opportunities for energy exports by decreasing the domestic use of fossil fuels. It should be of particular interest for Russia to make use of its large potential energy efficiency and renewable energy in order to be able to increase its energy exports. Secondly, Russia can attract significant investment by using mechanisms in the global climate regime that require increased production of renewable energy.

Thirdly, fossil fuels are exhaustible resources, whereas renewable energy sources are not. This means that developing renewable energy sources will be necessary sooner or later. Adjusting to future energy systems is important for all Kyoto Protocol signatories, even ones like Russia that are prone to put economic development before climate obligations. Being proactive in this development will undoubtedly be an advantage. Fourthly, Russia's vast size means that renewable energy solutions are the most economically viable in certain isolated areas such as the Northwest. And finally, Russia has several competitive advantages linked to its natural resource base and its strong tradition of research in the natural and technological sciences. Russia has the advantage of its geographic size and the variation in its climate and terrain, giving it the potential to develop virtually any kind of renewable energy. There are nevertheless difficulties in ascertaining the exact potential of the various resources, and estimates diverge. Contrasting Russia's potential for renewable energy with installed capacity reveals considerable scope for expansion.

Energy efficiency is also highly relevant to the prospects of making Russia's energy sector more environmentally friendly. Russia is one of the world's least efficient countries in terms of the amount of energy it uses. In contrast to the Russian leadership's seeming indifference to renewable energy, improving energy efficiency is recognized as important in the country's energy strategy. Both industrial and residential customers alike are wasteful in their energy use. An important reason for this is how energy has been priced. The prices of energy have not reflected production costs, and ever since the Soviet period, access to inexpensive energy has been granted to both industry and residents. In a market economy, this is not sustainable, and energy prices are set to increase. We argue that the Russian comparative advantage in renewables, along with the extreme inefficiency in current energy use, provides Russia with considerable potential for contributing to greener energy use globally.

Conditions for renewable energy in Russia

Chapter 2 on the reform of the Russian electricity sector drew up some background factors that help explain the limited exploitation of renewable energy sources in Russia. We see the continued existence of subsidized prices for domestic use of natural gas as being the main impediment to renewable energy development in Russia. This is not changed by the reform of the Russian electricity monopoly RAO UES. Although the reform may improve some factors, it is still unclear whether it will be able to create the conditions for wholesale market competition, which is in fact the goal of the reform. As regional monopoly-like situations are expected to continue to exist, this is likely to remain an impediment to large-scale development of renewable energy in Russia in the short to medium term.

Much will depend on not just the design of the reform, but its implementation. Examples from countries that have had a more developed market economy, a better regulatory framework in place and more transparent business environments than Russia when embarking upon electricity sector reforms clearly show the pitfalls in this area; Moscow risks repeating some significant errors. The likelihood that the reform will create significant levels of market power and even monopoly power under peak-load conditions in certain regions is seen as a serious hindrance with regard to development of renewable energy since difficulties are created for entry of newcomers to the market. This calls for establishing truly independent regulatory mechanisms, and indicates the necessity of building incentives and the right institutions to enable Russia to realize its potential in renewable energy sources and energy efficiency.

Since there is no proper Russian strategy for the development of renewable energy sources, it seems likely that in the short to medium term they will be overshadowed by Russia's petroleum resources, which are central to the country's

Energy Strategy of 2003. The Russian government needs to develop a clearer strategy and establish institutions that can ensure the development of renewable energy sources and the implementation of energy efficiency measures at all levels. International actors could play a role in getting Russian authorities to formulate such a strategy. The UN Climate Change Conference to be held in Copenhagen in December 2009 and the results of the negotiations to bring the Kyoto regime forward could provide a framework for international actors to put pressure on Russia with regard to formulating a more environmentally friendly energy policy.

Markets for renewable energy in Russia

Niche markets like remote Northern settlements or *dachas* can act as levers for the expansion of renewable energy. Especially replacing the expensive and old-fashioned transport of diesel to distant Northern settlements can be a way of showing the possibilities within renewable energy and developing important know-how without having to compete with subsidized natural gas. Avoiding this harsh competition, which may still be the main impediment to renewables development in Russia for years to come, can be crucial in the early phases of establishing a new energy sector. Remote settlements could emerge as one of the first realistic market niches for the profitable implementation of renewable energy in Russia, while waiting for better framework conditions in the rest of the country. As such, they could function as a testing ground for renewable energy, preparing the country and local and foreign actors for future expansion in this sector.

However, any introduction of renewable energy in remote locations should be done as part of a large-scale state-sponsored programme, in order to ensure economies of scale. To succeed, such projects will need the full support of the local and central authorities. Joint projects in this niche market could be counted as a joint implementation (JI) projects under the Kyoto Protocol and thus benefit the climate accounts of the Western partner country.

Matching Nordic and Russian comparative advantages

The case study of the Nordic countries' efforts within renewable energy in Russia offers useful insights for international actors seeking to establish renewable energy partnerships with Russia. It is our hope that the case study as well as the policy recommendations offered later in this chapter will help others to avoid these pitfalls.

Our view on cooperation as being about mutual benefit in science and innovation through complementarities has shaped our analysis of the existing Nordic–Russian cooperation within renewable energy. Apart from the possible market niches in replacing the Northern Freight system with renewable energy, we have found that international actors over-focus on setting up renewable energy generation capacity in Russia for the Russian market. At the same time, the potential that lies in cooperating with Russian researchers on basic science is often (though not always) ignored by foreigners. Three aspects of the existing Nordic–Russian cooperation within renewable energy are not the best way of maximizing the results of such cooperation.

First of all, existing Nordic-Russian cooperation in renewable energy has focused on setting up generation capacity for renewable energy in the Russian electricity market. Given the challenges that exist with regard to gaining grid access for independent producers, and the subsidized gas which makes it very difficult for other energy sources to compete, this might not be the most fruitful way of advocating renewable energy on the Russian agenda, nor is it the most desirable way of spending Nordic funds – particularly if the goal is to spur independent Russian efforts within renewable energy development. This approach implies an unfavourable logic of Western aid to 'underdeveloped' Russia and fails to acknowledge and benefit from the country's strong traditions in natural science.

Second, we have noted that much Nordic–Russian cooperation within renewable energy is geographically centred on Northwest Russia and areas in the proximity of the Nordic countries. Knowing what we have learned about the regional monopolies from the context chapter, this may not be the best region to invest Nordic resources – geographic factors notwithstanding.

Third, Chapter 7, which presented ten Nordic experiences from Nordic–Russian cooperation, indicated that the main foci of Nordic actors in Russia are wind and bioenergy. While Russia undoubtedly has great potential for the development of both of these energy sources, this narrow Nordic focus misses some of Russia's main scientific and technological strengths.

A more appropriate way of starting international cooperation with Russia would be for foreign partners to search out high-level competence in areas of basic science that are central to renewable energy. Coupling this with international expertise in management, commercialization and marketing might allow both parties to maximize the benefits of cooperation while drawing on their own strengths.

Particular areas where we believe there are large unrealized opportunities for international cooperation are within hydrogen and photovoltaics. The Russian solar power industry combines relatively high efficiency with low costs. International actors within solar energy can therefore benefit greatly from looking into cooperative ventures with Russian actors in this field.

The Russian scientific-educational system

In order for international actors to grasp Russia's long history of research and vast landscape of scientific of education institutions, we have included an overview of the country's scientific-educational system. In doing so we have singled out organizations that fund research in Russia, which may be of use to Western research funding organizations that wish to set up joint calls.

The collapse of the Soviet Union brought several challenges to the scientific-educational system. The economic chaos of the 1990s had severe consequences, and there has been a double brain drain from Russian science: from research to other sectors of the Russian economy, and from Russia to other countries. The decline in funding and the lack of opportunities for young scientists are equally worrisome developments. Today the outlook has improved slightly with an increased government focus on science and some of the country's largest companies maturing, and prospects for their increased interest in investments in research, given the right incentives. In recent years the research and development share of Russian GDP has increased, and the disbursement of state funding is now more timely and predictable than during the difficult 1990s. State funding, however, is taking a different shape now compared to the Soviet period. There is a movement towards making government research and development funding more transparent, target-oriented and efficient, and more competition is being introduced.

We have compiled an overview of some key Russian funding institutions for research that are of relevance for renewable energy development and that could be possible partners for international actors interested in co-financing. We have also prepared a ranking of excellent Russian research and education institutions in renewable energy, which reveals a surprisingly large number of institutions engaged specifically in renewable energy, often with a longer track record than one might expect.

Innovation

The economic growth experienced by Russia during the eight years preceding the financial crisis was largely dependent on high commodity prices. Focusing on research and innovation would help to foster new industries, increase productivity and diversify the Russian economy. Russian innovation indicators have remained disappointing, in spite of the country's potential.

Russian research and development is still primarily financed by the state. Innovation is rather low in the private sector, which has focused on imitation rather than research-based innovation. Poor communications between the public and private sectors have affected the levels of commercialization. With most research and innovation being state-funded, researchers in the public sphere have generally had little incentive to worry about the commercial applications of their work. A second major constraint on commercialization is Russia's weak intellectual property rights framework. The educational system may possibly also be having an effect on commercialization by failing to prepare students properly for market-oriented jobs.

International actors aiming to carry out commercialization in cooperation with Russian partners must therefore make clear both the benefits to be achieved from commercialization and the opportunities available. The Russian government appears to be aware of the need to reform its innovation policy, and has launched a strategy for the development of science and innovation to 2015 that includes both institutional reform and targeted initiatives.

Russia and the Kyoto Protocol

Many Russian scientists, officials, bureaucrats and politicians remain sceptical about the causal link between human-induced greenhouse gas emissions and climate change. Climate change issues rank correspondingly low on the Russian political agenda. Russia has nonetheless committed itself to the Kyoto Protocol, and should be encouraged to meet its international obligations in this field. If international actors could help to change Russian climate perceptions and policy, this would effectively contribute to the development of renewable energy in Russia and cross-border cooperation with Russia in this field. While acknowledging the scholarly disagreement but emphasizing that Russian scientists are among a dwindling minority of sceptics, international actors should try to help Russian actors to understand and realize the commercial benefits that they could achieve by engaging more actively with the Kyoto Protocol – for example, by showing an interest in emissions trade and Joint Implementation (JI) projects in Russia.

Due to its position as one of the most-energy intensive economies in the world, Russia offers a prime opportunity for Joint Implementation (JI) projects under the Kyoto Protocol. This enables EU countries to fulfil their commitments to reducing emissions under the Kyoto protocol, while at the same time establishing a presence in the potential Russian renewable energy market. Of the 163 JI projects currently underway in the world, 109 are located in Russia, most of them initiated by actors in the EU countries.

Policy recommendations

The commendations below are based both on this book and on the broader research of which the book is a product. (See the sections on methodology and data in the introductory chapter and in the appendices for further information.) The recommendations are grouped thematically and in order of priority.

Funding and coordination

Our first three recommendations are particularly relevant for foreign financial institutions seeking to engage with Russia's renewable energy sector. They focus on how international institutions can maximize the effect of their funds in cooperation with Russia, and are based on our analysis of existing cooperation as well as of the potential for further cooperation.

Coordinate and pool funding   A weakness of some international funding is that it is spread too thinly and between too many actors to serve as the main financing for serious research projects. This is particularly evident in the case of Nordic–Russian co-operation. This can be an obstacle especially when dealing with Russian actors, who will rarely be familiar with smaller international actors like the Nordic institutions in the first place. Better coordination among international funding bodies will also make it easier for prospective Russian partners to see and understand the opportunities. For further information, see Chapter 6, 'Nordic–Russian Cooperation on Renewable Energy'. See Chapter 5, 'EU–Russia Science and Energy Cooperation', for more on EU projects in the Russian energy sphere.

Top-down or bottom-up calls are fine   Several ideas to guide potential calls or instruments are presented in the recommendations below. These are based on our relatively comprehensive review of renewable energy in Russia, and should therefore provide a sufficient basis for top-down calls of limited thematic scope. If one moves away from the suggested topics and complementarities, a bottom–up approach might be considered. A bottom-up approach would also be one way to extend this policy survey. Asking foreign and Russian researchers to offer their own ideas for the best joint projects could yield more comprehensive information about what they see as the best areas of cooperation. For an overview of Nordic–Russian projects see Chapter 6, 'Nordic–Russian Cooperation on Renewable Energy'.

Don't worry too much about joint financing   Foreign organizations should not be overly worried about whether their Russia-oriented instruments are jointly financed by the Russian side or not. In some cases this should be possible to achieve, in others it may be difficult. In general it will be more difficult for a small institution from, say, a Nordic country to find Russian co-financers than for a big Chinese or Japanese company, which will attract more attention and prestige on the Russian side. Joint financing of, for example, projects within the Seventh Framework Programme should therefore not be taken at face value as evidence that all Russia-oriented projects now can or should be jointly financed. Also, Russia and the Baltic countries are different, so success in achieving joint financing of projects with the Baltic countries should not be interpreted as evidence that all Russian projects should also be jointly financed. See Chapter 5, 'EU–Russian Science and Energy Cooperation', for more information about what the EU and Russia are doing, and

Chapter 6, 'Nordic-Russian Cooperation on Renewable Energy', for information about existing Nordic–Russian cooperation.

Benefit from Russian strengths

As outlined in the introduction chapter, our understanding of cooperation is that it should ideally be about mutual benefit in science and innovation. This is why the following seven recommendations focus on how international actors can benefit from Russia's existing strengths in cooperation on renewable energy. These recommendations are relevant to all actors – policy-makers, financial institutions, researchers, entrepreneurs – seeking to engage in cooperation with Russian partners within renewable energy.

Work with these co-funders   The following are among the organizations suitable for doing joint calls or instruments in cooperation with Russia: the Russian Foundation for Basic Science, New Energy Projects (NIC-NEP), RusGidro. For more information about these and other Russian research funding organizations, see the section on research funding in Chapter 3, 'The Knowledge-base for Renewable Energy in Russia: Education, Research and Innovation'.

Work with the best institutes, build a brand   In order to foster the best projects with Russian partners, it is important for the funding organization to build a long-term reputation and make its calls highly visible and intelligible. This is relevant for international financing bodies. For a ranking of the top institutions within education and research on renewable energy, complete with internet addresses, see the section 'Top Russian institutions' in Chapter 3. The institutions in this ranking appear highly relevant for foreign funding organizations or research institutions looking for Russian project partners.

Seek out complementarities   Russian actors tend to be relatively strong in basic science but weaker on commercialization. Funding for international projects with Russia could aim to match Russian actors with strong competence in basic science and foreign actors with expertise in commercialization and marketing. This would enable international actors interested in engaging in the renewable energy sector in Russia to benefit more from existing Russian strengths. For international financial institutions this would increase the effectiveness of invested funds. See Chapter 3 – 'The Knowledge-base for Renewable Energy in Russia: Education, Research and Innovation' – for more on Russian strengths in natural science and a list of the top research and education institutions within renewable energy in Russia. See Chapter 6 – 'Nordic-Russian Cooperation on Renewable Energy' – for a discussion of existing Nordic cooperation efforts within renewable energy in Russia.

Focus on basic science in Russia   International actors seeking to get involved in cooperation with Russians on renewable energy tend to focus on setting up generation capacity for renewable energy for the Russian electricity market. A more appropriate angle might be to search for high-level competence in areas of basic science that are central to renewable energy. Two areas where Russian science has traditionally been strong and where its researchers and companies could contribute significantly to international projects are solar power and hydrogen technologies. These are also two areas that are currently receiving considerable attention around the world. See Chapter 6 on 'Nordic–Russian Cooperation on Renewable Energy', and particularly the section on comparative advantages and complementarities, for a discussion of how this cooperation can benefit more from utilizing Russian strengths. See Chapter 4 for an analysis of Russia's solar power sector.

Aim for exports from Russia   While acknowledging the potential of Russia's natural resource base and large population, international actors should be aware of the current obstacles to producing and selling renewable energy in Russia. At present it may make more sense to engage in projects oriented towards the export of materials, equipment and/or energy *from* Russia, than in the production of renewable energy for the Russian market. On the other hand, although establishing renewable energy production in Russia may not yet be profitable in the short term, it could be worth doing for actors seeking to position themselves for the long term. In the long term Russia could prove to be an exciting market for renewable energy. See Chapter 6, 'Nordic–Russian cooperation on renewable energy', and Chapter 7, 'Working on Renewable Energy in Russia: Ten Experiences'. See also Chapter 2, 'Russian Energy Markets: Liberalization and Niches for Renewable Energy', for an analysis of obstacles to renewable energy in the Russian power sector.

Cooperate for mutual benefit   This book is oriented towards cooperation that is *mutually* beneficial. Should one want to engage in cooperation of a more aid-oriented nature, the most productive areas are those related to changing policy in order to facilitate growth in the use of renewable energy. See, for example, the recommendation below on influencing Russian climate policy. Transferring Western technology for windmills, small hydro, etc. makes little sense as long as the regulative framework and attitudes are not in place to implement such technology. For more information on Russian climate politics, see the section 'Russia and the international climate regime' in the Introduction.

Aim for niches in the North   For actors who still wish to install renewable energy production capacity in Russia, a logical choice would be to focus on projects that supply the remote towns and settlements currently served by the Northern Freight subsidy system. This is the most promising market niche for renewable energy in Russia today due to its isolation and the high costs of transporting fuels there. Increased international involvement may help the Russian authorities to recognize this. Joint projects in this niche market should engage both the Russian central authorities and the relevant local authorities. They could possibly be counted as a joint implementation projects under the Kyoto Protocol and thus benefit the climate accounts of the foreign partner. See the section on markets for renewable energy in Russia in Chapter 2 for an analysis of this possible niche.

Energy policy matters

We believe that energy policy matters for the outcome of international cooperation with Russia on renewable energy. The following three recommendations focus on how foreign actors can work to improve Russian renewable energy policy. Policy-makers and financial institutions are significant in influencing the shape of Russia's international cooperation with regard to policy issues. In showing an awareness of how energy policy matters, researchers or companies cooperating with Russian partners can also play a significant role in influencing Russian energy policy.

Work to change Russian climate policy   Many Russian scientists, officials, bureaucrats and politicians remain sceptical about the causal link between human-induced greenhouse gas emissions and climate change. Climate-change issues rank correspondingly low on the Russian political agenda. Russia has nonetheless committed itself to the Kyoto Protocol, and should be encouraged to meet its international obligations in this field. If international actors could help to change Russian climate perceptions and policy, this would effectively contribute to the development of renewable energy in Russia and international cooperation in this field. Here it would be important to:

- acknowledge that there exists scholarly disagreement between most Russian and international climate scientists
- recognize the historical strength of Russian climate science, especially in marine and atmospheric temperature measurement
- seek to make it clear that Russian climate scientists are nonetheless among a dwindling minority of sceptics
- help Russian actors to understand and realize the commercial benefits that they could achieve by engaging more actively with the Kyoto Protocol – for example, by showing an interest in emissions trade and JI projects in Russia.

This policy recommendation is particularly relevant for international research institutions within climate science, as well as for financial institutions wishing to finance research on climate issues between foreign and Russian partners or JI projects in Russia. It is also relevant knowledge for actors involved in negotiating a follow-up to the Kyoto Protocol and policy-makers involved in dialogue with Russian counterparts. See the introductory chapter for a brief description of Russian climate policy and lack of climate action on the part of Russian policy-makers, and Chapter 5 – 'EU–Russian Science and Energy Cooperation' – for more on the role of the Kyoto Protocol and JI projects in EU–Russia relations.

Support an improved Russian strategy   Russia lacks a clear and operationalized strategy for the development of its renewable energy sources. International actors and institutions should promote the formulation of such a strategy on behalf of the Russian authorities. See Chapter 2 – 'Russian Energy Markets: Liberalization and Niches for Renewable Energy' – for more on the Russian energy strategy.

Geographical considerations

Finally, we have found significant geographical bias in the case of Nordic–Russian cooperation, a problem that may also extend to other countries dealing with Russia. The next two recommendations focus on how international actors can maximize the benefits of engaging with Russia by broadening the geographic scope of their activities. This is relevant for actors from the private and public sectors alike.

Don't stare blindly at your part of Russia   Nordic–Russian collaboration in research has been oriented almost exclusively towards Northwest Russia.[2] Foreign involvement in Russia is subject to normal mechanisms of supply and demand. Due to the proximity of the Nordic countries and the rest of Western Europe, there has been an oversupply of opportunities to cooperate with Western partners in Northwest Russia – in renewable energy, energy efficiency and other areas. This is a relevant point in an area like research and particularly in a centralized country like Russia, where some of the most interesting potential partners are likely to be located in Moscow and other places outside Northwest Russia. International actors should thus ensure a broad geographical scope for their collaboration. See Chapter 3 – 'The Knowledge-base for Renewable Energy in Russia: Education, Research and Innovation' – for an overview of the Russian research system and a ranking of Russian institutions within education and research on renewable energy. See Chapter 6 – 'Nordic–Russian Cooperation on Renewable Energy' – for an overview of existing Nordic–Russian cooperation mechanisms.

Consider the Urals, South and Siberia   It is still not clear to what extent and in which areas the ongoing liberalization of the Russian electricity sector will create competition in electricity generation. However, some regions will have better conditions for competition. The Volga Region, the Central Region and the Northwestern Region are likely to be the worst cases of regional quasi-monopolies in electricity generation. Promoting renewable energy may be easier in the Urals, the South and perhaps Siberia, where more open markets are expected to develop. Again this is an indication that international actors should not focus blindly on the part of Russia that is geographically closest to them. This applies to actors from the private and public sectors, and business and research alike. See Chapter

---

2   Cf. Aasland, Aadne. *Development in Research: An Outline of the Science Systems in Russia and the Baltic States.* Oslo: Nordforsk, 2007. http://www.nordforsk.org/_img/ nordforsk_pb1_web.pdf [accessed 12 May 2008].

2 – 'Russian Energy Markets: Liberalization and Niches for Renewable Energy' – for a discussion of the introduction of competition in the Russian electricity market. See Chapter 6, 'Nordic–Russian Cooperation on Renewable Energy', for an overview of the complementary interests of the various Nordic countries and different parts of Russia.

Offer funding, equipment and relationships   Reviews of previous funding for research cooperation with Russia show that it is often the Russian researchers who initiate cooperation, and when they do, they may be particularly attracted by tax-free grants, duty-free imports of equipment, opportunities to work with highly advanced equipment otherwise unavailable, and the prospects of building long-term relationships with international partners – in addition to access to funding in general. Instruments aimed at facilitating cooperation with Russia should try to offer these elements in order to attract the best partners.

Russian renewable energy 2025: Scenario sketches

Having analysed the current potential for renewable energy cooperation, we now look to the future. In this section we briefly draw up three possible scenarios for the development of Russia as a global renewable energy actor in a medium-term perspective. These scenarios are not intended to be read as scientific predictions of what will happen, but to represent a range of possible developments that are at least plausible. The purpose of these scenarios is to help international actors interested in cooperating with Russian actors on renewable energy to think about possible future trajectories of Russia's renewable energy sector, including both optimistic and more sober possibilities. They are illustrated in Figure 8.1.

Scenario stories are built to illustrate possible situations at a specific moment in time as a result of different developments or actions. A picture with three scenario stories illustrates a scope of possible developments. The stories can be read as possible developments, but also be used as analytical tools to understand how today's decisions can shape the future, or what possible futures should be taken into account in taking decisions today.

As the two main drivers for future developments, we have chosen (1) government policy and (2) the business culture of Russia's large corporations. We chose to focus on big business because there is no doubt that many Russian SMEs are ready and eager to spearhead renewable energy: the question is whether they will be hindered or supported by the larger companies and the state. The government may be more passive or more proactive; Russian companies may continue to operate according to a chronically short-term, grab-and-run perspective, or they may mature and start investing for long-term growth.

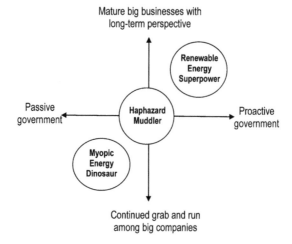

Figure 8.1   Scenario cross

In order to illustrate how an unlikely, unpredictable, but nonetheless possible event can have the potential to drastically change the country's role in the global renewable energy sector, we have also thrown in a 'wild card'.

The scenarios are based on a broad review of the data gathered for this book, in particular our impressions from the 98 interviews listed in the Appendix. These scenarios particularly draw on the following points made elsewhere in the book:

- the importance of big business and government regulation (Chapter 2)
- the importance of support mechanisms for renewable energy, in particular legal amendments providing advantageous tariffs and facilitating connecting to the electricity grid (Chapter 2)
- Russia's big natural resource base in hydroelectricity, geothermal energy and bio-energy outlined in the section 'Why renewable energy in Russia?' in the introductory chapter
- Russia's strong scientific position in photovoltaics (Chapter 4)[3]
- the many Russian excellent institutions of research and education, some of which have long traditions in renewable energy (Chapter 3)
- the need for a Russian strategy for innovation-based business development (Chapter 3).

These are a selection of the most interesting and/or promising aspects of Russia's renewable sector uncovered in this book, and function as the drivers of

---

3   Presentation by Indra Øverland at NUPI conference *Renewable Energy in Russia: How can Nordic and Russian actors work together?*, Oslo, Norway, 8 May 2008.

our scenarios. Whether the country can manage to realize its renewable energy potential will depend on whether it can realize the full potential of these factors. Ultimately, this also has serious implications for any Western actors wishing to get involved in the Russian renewable energy sector.

Scenario 1: Renewable superpower

In this scenario, global energy demand continues to rise, fuelled by the growth of key emerging players such as India and China. At the same time, Russia is using its hydroelectric, bioenergy and geothermal potential for the production of input factors for the expanding global renewables economy (hydrogen, silicon, etc.). This is supplemented by rapid technological development and industrial expansion in the areas of solar power and bioenergy.

Russia developed a strategy for the development of renewable energy sources in 2015. Through a deliberate strategy towards the goal of tripling renewable energy production by 2025, Russia has encouraged large private sector investment in renewables through tax rebates, and innovation is blossoming. Russia seems to be nurturing several world-class firms oriented towards the renewable energy sector, and investing significantly in further research and development. In an about-face in 2011, Gazprom started pouring capital into renewable energy. Much of this funding was wasted, but a few success stories were spun off in the big break-up of Gazprom assets in the later 2010s. Indeed Russian firms are now among the undisputed leaders in solar and geothermal energy, and hydrogen production and fuel cells.

WTO membership is attained in 2011 in return for phasing out subsidies to energy-intensive domestic industries over an eight-year period, whereas subsidies for the energy used by the general population are phased out over ten years. This, along with Russian adherence to the Kyoto Protocol, leads to a significant increase in energy efficiency, lower consumption and the expansion of renewable energy. The sum of these factors contributes to unprecedented growth, rises in income and a considerable strengthening of the country's international status.

Scenario 2: Myopic energy dinosaur

Energy demand continues to grow both inside and outside the country, but in this scenario Russia's energy industry stagnates due to lack of reform and investment. Oil and gas production gradually decrease as Russia fails to invest and develop new fields. Most international oil companies leave, taking their much needed technology with them.

Russia relies on petroleum exports to keep its economy going. These account for most of the government budget and continue to subsidize domestic energy intensive industries. The result is a bad case of 'Dutch disease' as the exchange rate rises and the economy has failed to diversify. Manufacturing and other sectors stagnate, unable to compete on the global market due to the appreciation of the rouble. At the same time the long-term subsidies for energy intensive industries have led to substantial inefficiencies, stunted innovation and even more reliance on subsidies (especially in light of the steadily appreciating rouble). In this context, renewable energy has absolutely no chance to develop. The Russian authorities fail to establish any meaningful support mechanisms for renewable energy, and the legal mechanisms setting out support rates for renewable energy and the rules on connecting renewable energy supplies to grids finally produced in 2011 prove ineffectual. At the same time, fossil fuels and nuclear power are in practice heavily subsidized.

In this scenario, public service reform is virtually non-existent; corruption is rampant and bureaucratic and institutional inefficiencies remain in place. Spending on research is dwindling, partly the result of falling oil production which has led to a reduced state budget and partly due to continued bandit capitalism among Russian companies, which are unable to attract long-term investors because of the uncertainty surrounding property rights, political intervention and taxation. Renewable energies have been ignored, and the huge potential of the renewable energy industry in Russia lies dormant. Foreign actors have instead flocked to other emerging economies which have followed China's example and created significant incentives for investment in the renewable energy sector. Foreign companies remain wary of Russia due to its continuing nationalization of private companies, widespread antipathy towards foreign capital among the population and an unwelcoming tax regime.

Scenario 3: Haphazard muddler

There has been some success with social and economic reforms but institutional and structural weaknesses abound. The field of renewable energy has received attention, but this has largely been rhetorical, and reform in this area remains particularly hollow. Funding for research and development continues to be dominated by the state, and most funding allocations are focused rigidly on institutions rather than project-based or competitive funding. A few renewable energy areas have received a trickle of funding and individual successes have been noted. These have mostly been in fields that draw heavily on Soviet scientific traditions, such as membranes for hydrogen fuel cells. The main progress in the areas of renewables comes in the form of foreign actors making significant investments. Their relationship with the government is bumpy, with the government at first welcoming these actors and providing some incentives – but then becoming less cooperative and more demanding as the photovoltaic and hydrogen sectors start booming, and a discussion ensues about taxation issues. Those foreign companies that aim for scientific innovation based on traditionally strong fields in Russian research and subsequently move people and results out of the country are the most successful. A few local enthusiastic researcher-entrepreneurs also do well, but not thanks to support from the government or big business.

In this scenario, the Russian economy has diversified somewhat as subsidies to energy intensive industries are slowly phased out and Russia achieved accession to the WTO in 2012. However, much structural inefficiency remains, and oil production has dropped due to a lack of investment and ailing infrastructure. Meanwhile there has been considerable success in other sectors, notably the IT industry. But in renewable energy, Russia fails to fulfil its potential.

Wild card

In this final scenario, three Russian physicists receive the Nobel Prize for their achievements in the area of hydrogen research. The three academics, professors at the Novosibirsk State Technical University, have developed a simple, cost-effective way of transforming electricity into hydrogen. This scientific process, combined with progress in research on membranes for fuel cells at the IOFFE Institute that allows the safe and compact storage of the hydrogen and with the country's large hydroelectric and nuclear resources, facilitates Russia becoming the spearhead of the global hydrogen economy.

With the Russian state owning the patent rights to these new technologies, the benefits for the Russian economy will be tremendous. Oil prices have halved during the past four months, and are expected to keep dropping as this new technology is commercialized for large-scale production. Local markets have boomed since the news, after years of rising energy prices fuelled by various supply constraints. The economic windfall accruing to Russia is assumed to make it the world's third largest economy after China and India.

Appendices

Table A.1   Russian funding institutions for research

	Type of organisation	Relevance to renewable energy	Web address
Russian Foundation for Basic Research	Non-commercial government organisation. Receives a fixed share of the Russian federal budget for basic research and technological development. In 2005 this share was 6%.	Support to initiative-based research. Funding for basic science.	http://www.rfbr.ru/eng/
Ministry of Education and Science	Ministry	Broad range of funding activities.	http://www.mon.gov.ru/
Russian Academy of Sciences	Non-commercial independent academic organisation	Broad range of funding activities.	http://www.ras.ru/index.aspx?_Language=en
Federal Agency of Education	Federal Agency	Broad range of funding activities.	http://www.ed.gov.ru/
Federal Agency of Science and Innovation	Federal Agency	Broad range of funding activities.	http://www.fasi.gov.ru/
Ministry of Natural Resources	Ministry	Broad range of funding activities.	http://www.mnr.gov.ru/
New Energy Projects (NIC-NEP)	Private–public partnership. Distributes 30–40 mill annually	Coordination of research on hydrogen technologies. Research on hydrogen and fuel cells. Commercialisation of wind and solar energy.	http://www.nic-nep.com/
National Centre for Monitoring of Innovation Infrastructures	Information service relating to the development of national innovation systems.	Monitoring and coordination of innovation activities in Russian regions.	http://www.miiris.ru/
RusGidro	The biggest private-sector company involved in renewable energy in Russia.	Broad engagement in renewable energy, especially hydro-electric, but also tidal and wind. Exact structure of company not clear, because it was only recently formed after RAO UES dissolution.	http://www.gidroogk.ru

The following table provides an alphabetical list of all of the institutions included in the ranking in Chapter 3. Where we found them, English-language web pages have been given preference over Russian ones.

Table A.2   Table of all ranked institutions

English title	Russian title	Web site
All-Russian Institute for the Electrification of Agriculture (VIESH)	Vserossiyskiy nauchno-issledovatelskiy institut elektrifikatsii selskogo khozyaystva (VIESKh)	www.viesh.ru
Bauman Moscow State Technical University (MSTU)	Moskovskiy gosudarstvennyy universitet imeni Baumana (MGTU)	www.bmstu.ru
Boreskov Institute of Catalysis SB RAS Novosibirsk	Institut kataliza imeni Boreskova SO RAN Novosibirsk	www.catalysis.ru
Buryat State Agricultural Academy	Buryatskaya Gosudarstvennaya Selsko-Khozyaystvennaya Akademia	http://www.bgsha.ru/eng/index.html
Central Aerodynamics Institute	Tsentralnyy aerodinamicheskiy institute (TsAGI)	www.tsagi.ru/eng/
Dagestan State University	Dagestanskiy gosudarstvennyy universitet	www.dgu.ru
Dubna International University of Nature, Society and Humanity	Mezhdunarodny universitet prirody, obshchestva i cheloveka 'Dubna'	http://www.uni-dubna.ru/
Energy Strategy Institute	Institut energeticheskoy strategii (Bezrukikh et al.)	http://www.energystrategy.ru/
RusHydro Scientific Research Institute of Energy Construction	Nauchno-issledovatelskiy institute energeticheskikh sooruzheniy RusGidro	www.niies.ru
Goryachkin Moscow State Agro-Engineering University	Moskovskiy gosudarstvennyy agroinzhenernyy universitet im. Goryachkina (MGAU)	http://www.ucheba.ru/vuz/first-5847.html
International Science and Technology Centre (ISTC)	Mezdhunarodnyy nauchno-tekhnicheskiy tsentr (MNTTs)	www.istc.ru
Ioffe Physico-Technical Institute of the Russian Academy of Sciences	Fiziko-tekhnicheskiy isntitut imeni Ioffe Rossiyskoy akademii nauk	www.ioffe.ru
Ivanovo State Power University	Ivanovskiy gosudarstvennyy energeticheskiy universitet	http://www.ispu.ru/portal/index.pl?iid=13692
Kazan State Energy University	Kazanskiy gosudarstvennyy energeticheskiy universitet (KGEU)	http://www.kgeu.ru/
Kostyakov All-Russian Institute of Hydrology and Irrigation	Vserossiyskiy nauchno-tekhnicheskiy institute gidrotekhniki melioratsii imeni Kostyakova	www.vniigim.ru
Krzhizhanovsky Power Engineering Institute (ENIN)	Energeticheskiy institut imeni Krzhizhanovskogo (ENIN)	http://www.mtu-net.ru/lge/english/inden.htm
Lomonosov Moscow State University (MGU)	Moskovskiy gosudarstvennyy universitet imeni Lomonosova (MGU)	www.msu.ru
Mari State Technical University	Mariyskiy gosudarstvennyy tekhnicheskiy universitet (MarGTU)	http://www.marstu.mari.ru:8101/marstu-en.html
Moscow Power Engineering Institute (MPEI)	Moskovski energeticheskiy institut (MEI)	http://www.mpei.ru/lang/eng/main/about/general/general.asp

Table A.2   Continued

English title	Russian title	Web site
Moscow Aviation Institute (MAI)	Moskovskiy aviatsionnyy institute (MAI)	www.mai.ru/english/
Moscow State University of Ecological Engineering	Moskovskiy gosudarstvennyy universitet inzhenernoy ekologii (MGUIE)	www.msuie.ru
Moscow State University of Railway Engineering	Moskovskiy gosudarstvennyy universitet putey soobshcheniya	www.miit.ru/engl/index.htm
Novosibirsk State Technical University (NSTU)	Novosibirskiy gosudarstvennyy tekhnicheskiy universitet (NGTU)	info.nstu.ru
St. Petersburg State Mining Institute	Sankt-Peterburgskiy gosudarstvennyy gornyy institute	http://www.spmi.ru/news/2/542
St. Petersburg State Polytechnical University	Sankt Peterburgskiy gosudarstvennyy politekhnicheskiy universitet (SPGPU)	http://www.spbstu.ru/english/index.html
Tomsk Polytechnical University (TPU)	Tomskiy politekhnicheskiy universitet (TPU)	http://www.tpu.ru/
United Institute of High Temperatures of the Russian Academy of Sciences	Obedinennyy institut vysokikh temperatur Rossiyskoy akademii nauk (OIVTRAN)	www.ihep.su
Urals State University	Uralskiy gosudarstvenny universitet (UrGU)	http://www.usu.ru/

Table A.3   Russian solar power manufacturers

Institute/company	Found.	Products	City	Address
Ioffe Institute	1918	Solar cells	St. Petersburg	194021 St. Petersburg, Politekhnicheskaya 26
All-Russian Institute for the Electrification of Agriculture (VIESH)	1930	27–33 W photoelectric modules and photoelectric power stations for small-scale use	Moscow	109456, Moscow, 1st Veshnyakovskiy Street 2, tel. (095) 170-16-65, fax (095) 170-51-01, e-mail: energy@viesh.msk.ru
AO VIEN (VIESH subsidiary)	–	–	Moscow	Fax 174-81-13, tel. 174-81-13
Kvant	–	11; 22; 33; 44; 50; 66; 80 and 110 W photovoltaic modules, all-round photoelectric power stations	Moscow	129626 Moscow, 3rd Mytishchenskaya Street 16, tel. (095) 287-98-28, fax (095) 287-18-71
NPF Kvark	–	3-60 W photoelectric modules	Krasnodar	–
1OAO Kovrovskiy Mechanical Factory	1940	Solar modules	Kovrov, Vladimir Oblast	601909 Vladimir Oblast, Kovrov, Sotsialisticheskaya Street 26, tel. (09232) 942-86, fax 30831, 34754
NPO Mashinostroeniya	–	–	Reutov, Moscow Oblast	Tel. 528-30-54, fax 302-50-90
FGUP NIIPP	1964	0.3 200 W photoelectric modules	Tomsk	634034 Tomsk, Krasnoarmeyskaya Street 99a, fax (382-2) 55-50-89, tel. (382-2) 48-81-59-48-82-12, e-mail opt@niipp.tomsk.ru

Table A.3   Continued

Institute/company	Found.	Products	City	Address
ZAO OKB Krasnoe Znamya Factory	1984	10; 15; 25; 30; 33; 35; 40; 45; 50; 55 and 60 W photoelectric modules	Ryazan	390043 Ryazan, Prospekt Shabulina 2a, tel. (0912) 53-84-03, 53-84-42, 53-85-39
AOOT Pozit Pravdinskiy Research Factory on Electricity	–	4.5; 5; 8; 9 and 10 W photoelectric modules with and without frame; photoelectric power stations	Pravdinskiy Village, Moscow Oblast	141290 Moscow Oblast, Pravdinskiy Village, Fabrichnaya Street 8, tel. (095) 584-62-52, 584-34-02, fax (095) 584-32-82
OAO Ryazan Metal Ceramics Instrumentation Plant (RMCIP)	1963	8-55 W photoelectric modules	Ryazan	390027 Ryazan, Novaya Street 55, tel./fax (0912) 44-19-70, www.rmcip.ru
OOO Solnechny Veter	1992	3-200 W photoelectric power stations for small-scale use	Krasnodar	350000 Krasnodar, Bazovskaya Street 69, tel./fax (8612) 55-22-86, 68-55-05, e-mail solwind@krasnodar.ru
OOO SOVLAKS	1991	12.5-20 W hard photoelectric modules based on amorphous silicon; 20-80 W photoelectric modules for roofs based on amorphous silicon	Moscow	129626 Moscow, Kulakov Alley 15, tel. (095) 287-97-58, fax (095) 286-35-67
OAO Saturn	–	10; 25 and 55 W photoelectric modules; 10; 100; 200 and 500 W photoelectric power stations; universal 0.06-10 kW pv power stations	Krasnodar	–
NPF Sun Energy	–	Eight different types of photoelectric power station	Moscow	129626 Moscow 3rd Mytishchinskaya Street 16, Building 60, tel. 287-96-36, 287-98-40, fax 287-67-97
3AO Telekom STV	1991	5; 10; 20; 22; 25; 30; 33; 35; 40; 45; 50 and 53 W photoelectric modules	Moscow / Zelenograd	103527 Zelenograd, Solnechnaya Ave. 1, tel. (095) 531-83-51, 532-90-36
AO Elma	–	5; 7; 10; 12; 30; 33; 35; 40; 45 and 50 W photoelectric modules	Moscow	

Peer review and conferences

The project proposal and this book have been peer reviewed by three prominent experts:

- Peter Lund, Professor in Advanced Energy Systems, Helsinki University of Technology
- Elena Merle-Béral, Regional Programme Manager, International Energy Agency
- Valery Kharchenko, Professor, Department of Renewable Energy, All-Russian Research Institute for the Electrification of Agriculture.

Table A.4 **CO_2 per capita in Russia and the Nordic countries, metric tonnes CO_2 per capita[1]**

	1992	1996	2000	2004
Norway	7,1256	9,1747	9,8936	19,0086
Finland	9,5047	12,1459	10,0011	12,5782
Russian Federation	13,2955	10,0333	9,9761	10,5393
Denmark	10,3284	12,7519	8,6761	9,8013
Iceland	6,9832	8,1851	7,6825	7,6103
Sweden	5,9315	6,0844	5,2429	5,894

The role of the peer reviewers has been to provide corrections and advice on how to proceed. Their input has been highly valuable for the project. However, their power has been limited and we have been at liberty to adhere to or disregard their suggestions, depending on our time and resource limitations. The contents of this book are therefore solely the responsibility of its authors. Project output has been presented at four international conferences, where it has received feedback from qualified audiences:

- Nordic Energy Research Policy Studies, Mid-term Review, Holmenkollen, Oslo, 3 December 2007
  - 'Presentation of project progress' (Øverland)
- North-West Russian Renewable Energy Forum, Murmansk, 1–2 April 2008
  - 'Markets for Russian Renewables' (Øverland)
- Renewable Energy in Russia: How Can Nordic and Russian Actors Work Together, Oslo, 8 May 2008
  - 'Russia's Hydrogen Sector' (Øverland)
  - 'Lost in Translation? Experiences and Perspectives from Nordic–Russian Cooperation on Renewable Energy' (Madsen)
  - 'Legal Framework: Conditions for the Development of Renewable Energy Sources in Russia' (Kjærnet)
- *Nordic Energy Policy Workshop*, Holmenkollen, Oslo, 1 December 2008
  - 'Russian Energy Research and Innovation: Prospects for Cooperation on Renewables and Efficiency' (Øverland)

The most important of these was the NUPI conference, with 57 participants and three presentations related to this book. Finally, we have received extensive feedback and input from Nordic Energy Research.

---

1 Source: UN Statistics Division. http://unstats.un.org/unsd/cdb/cdb_series_xrxx. asp?series_code=30248 [accessed 9 June 2008].

Interviewees and interlocutors

The following pages provide an overview of the people we have spoken to in connection with this project. With some individuals we have had little more than a fleeting, informal conversation. Others were subjected to fully structured interviews, while yet others were interviewed by telephone in order to cram in as many interviews as possible into the time-span of the project. Interviews were carried out on seven fieldwork trips:

Table A.5   Fieldwork

Location	Researcher	Date
St Petersburg	Øverland	12–13 November 2007
Moscow	Madsen	26–27 November 2007
Moscow	Kjærnet	5 February 2008
Moscow	Øverland	26 February–4 March 2008
Copenhagen	Madsen	12–13 March 2008
Murmansk	Øverland	1–2 April 2008
Stockholm	Madsen	13–16 April 2008

Key to codes in table on following pages

HK – Heidi Kjærnet
IO – Indra Øverland
NKM – Nina Kristine Madsen
GD – Grant Dansie

The interviewees in the table on the following pages are listed by order of date, to show the trajectory of our meetings and conversations.

Interview guide

1: Renewable energy (RE)

* How would you define renewable energy?
* What do you think of when discussing renewable energy?

2: Concept of Nordic–Russian cooperation

* From your perspective, *why* is Nordic–Russian cooperation on renewable energy interesting?

Table A.6   Interviewees

	Name	Position	Institution	Short informal conversation	Tel. interv.	Location	Date (d/m/y)	By
1	Björn Gunnarsson	Academic Director	School for Renewable Energy Science, Akureyri, Iceland			Helsinki	01.10.07	IO
2	Nina Lesikhina	Energy Project Coordinator	Bellona Murmansk			Oslo	23.10.07	IO
3	Michele Grønbech	Clean Energy Advisor	Russian Department, Bellona			Oslo	23.10.07	IO
						Oslo	14.03.08	NKM
4	Grigori Dmitriev	Vice President	World Wind Energy Association / Kola Science Centre			Oslo	23.10.07	IO
						Oslo	10.03.08	IO
						Murmansk	01.02.08	IO
5	Lars Helge Helvig	Senior Manager	Norsk vindenergi AS			Oslo	23.10.07	IO
6	Tatiana Mitrova	Head	Centre for International Energy Market Studies, Energy Research Institute, Russian Academy of Sciences, Moscow	X		Zurich	27.10.07	IO
7	Vyacheslav Kulagin	Deputy Director	Centre for International Energy Market Studies, Energy Research Institute, Russian Academy of Sciences, Moscow			Zurich	27.10.07	IO
8	Boris Novikov	General Manager	Innovation Journal	X		Petersburg	12.11.07	IO
9	Leonid Bobylev	Director	Nansen International Environmental and Remote Sensing Centre, St. Petersburg	X		Petersburg	12.11.07	IO
10	Philip Kazin	Senior Project Manager	St. Petersburg State University	X		Petersburg	12.11.07	IO
11	Alexei Bambulyak	Manager	Akvaplan-NIVA Russia	X		Petersburg	12.11.07	IO
12	Salve Dahle	Director	Akvaplan-NIVA	X		Petersburg	12.11.07	IO
13	Arne Grove	Director	Nordic Council of Ministers Information Office in Kaliningrad	X	X	Petersburg	12.11.07	IO
						Murmansk	02.02.08	IO
						Kaliningrad	07.03.08	NKM
						Oslo	08.05.08	IO
14	Boris Brygunov	President	Elektrosfera			Petersburg	13.11.07	IO
15	Iakov Blyashko	General Director	INSET JSC			Petersburg	13.11.07	IO
16	Viktor Elistratov	Chair	Renewable Energy Sources and Hydro-Power Engineering Department, St. Petersburg Polytechnical State University			Petersburg	14.11.07	IO
17	Vadim Dormidontov	Senior Banker, Power and Energy	European Bank of Reconstruction and Development, Moscow			Moscow	05.02.08	HK
18	Vitaly Yurchenko	Senior Research Fellow	Institute of Physics, University of Oslo		X	Oslo	23.02.08	IO

Table A.6 Continued

	Name	Position	Institution	Short informal conversation	Tel. interv.	Location	Date (d/m/y)	By
19	Pavel Bezrukikh	Assistant Director	State Institute of Energy Strategy	X		Moscow	27.11.07	NKM
						Moscow	05.02.08	HK
20	Tatiana Konik	Head	Information Services Division, Statistics of Russia		X	Moscow	26.02.08	IO
21	Inna Gritsevich	Coordinator of Project on Energy Efficiency	WWF Russia			Moscow	26.02.08	IO
22	Alexey Knizhnikov	Policy Officer	Oil and Gas Section, WWF Russia			Moscow	26.02.08	IO
23	Grigory Grigorievich	Responsible for publications	Izdatelsko-Analiticheskiy Tsentr Energiya			Moscow	27.02.08	IO
24	Irina Yazvina	First Deputy Director	Federal State Body Rosinformresurs, Ministry of Industry and Energy			Moscow	27.02.08	IO
25	Igor Tyukhov	Deputy Chair	UNESCO Chair of Renewable Energy and Rural Electrification, All-Russian Research Institute for the Electrification of Agriculture, Russian Academy of Agricultural Science, Moscow			Moscow	27.02.08	IO
26	Dmitriy Strebkov	Director	All-Russian Research Institute for the Electrification of Agriculture			Moscow	27.02.08	IO
27	Vladislav Larin	Editor	Department of Technology of the Academy of Sciences journal Energy: Economics, Technology, Ecology			Moscow	27.02.08	IO
28	Vladimir Vissarionov	First Deputy Director	Faculty of Non-Traditional and Renewable Sources of Energy, MPEI			Moscow	28.02.08	IO
29	Mikhail Sleptsov	Head	Department of International Cooperation, MPEI			Moscow	28.02.08	IO
30	Igor Bashmakov	Executive Director	CENEF Centre for Energy Efficiency			Moscow	28.02.08	IO
31	Irina Aksenova	Senior Energy Specialist	US Department of Energy, Moscow Office			Moscow	29.02.08	IO
32	Svetlana Frenova	Regional Manager	Renewable Energy and Energy Efficiency Partnership (REEEP), Regional Secretariat for Russia and FSU			Moscow	29.02.08	IO
33	Vladimir Karghiev	Manager	OPET CIS, Intersolar			Moscow	29.02.08	IO
34	Bladimir Kabakov	Head	Department of Marketing and Information (ENIN)			Moscow	03.03.08	IO
35	Boris Tarnizhevsky	Head of Department	(ENIN)			Moscow	03.03.08	IO

Table A.6   Continued

	Name	Position	Institution	Short informal conversation	Tel. interv.	Location	Date (d/m/y)	By
36	Sergei Fadeev	Vice Director	(ENIN)			Moscow	03.03.08	IO
37	Igor Podgorny	Manager of Energy Efficiency Project	Greenpeace			Moscow	04.03.08	IO
38	Nikolay Filatov	General Director	Energetika goroda			Moscow	04.03.08	IO
39	Maral Ovezova	Independent Economist	–	X		Moscow	04.03.08	IO
40	Nazar Suyyunov	Former Minister of Oil and Gas	USSR, Turkmen Socialist Republic	X		Moscow	04.03.08	IO
41	Natalia Grebennik	Advisor	Nordic Energy Research	X		Oslo	10.03.08	IO, HK, NKM
42	Inna Rudakova	Researcher	Department of Environmental Engineering, Saint Petersburg State Technological University of Plant Polymers			Oslo	10.03.08	IO, HK, NKM
43	Anna Kulikovskaya	Technical Director	Energy Saving Foundation of the Arkhangelsk Region			Oslo	10.03.08	IO, HK, NKM
44	Aleksei Shtykov	Director	Center of Excellence at Petrozavodsk State University	X		Oslo	10.03.08	IO, HK, NKM
45	Marina Isakova	Administration, Specialist in Tourist Department	Committee for physical culture, sport, tourism and youth politics, Leningrad region	X		Oslo	10.03.08	IO, HK, NKM
46	Raimonda Makrickaite	Projects Consultant	Lithuanian Innovation Centre	X		Oslo	10.03.08	IO, HK, NKM
47	Ilze Skrebele-Stikane	Project Manager	The Knowledge and Innovation System Department, Investment and Development Agency of Latvia	X		Oslo	10.03.08	IO, HK, NKM

Table A.6   Continued

	Name	Position	Institution	Short informal conversation	Tel. interv.	Location	Date (d/m/y)	By
48	Imre Mürk	Specialist	Ministry of Economic Affairs and Communications. Division of Innovation and Technology	X		Oslo	10.03.08	IO, HK, NKM
49	Yana Bocharova	Personell and Project Manager	Information Office of the Nordic Council of Ministers in Saint Petersburg	X		Oslo	10.03.08	IO, HK, NKM
50	Louis Skyner	Attorney and Head	Russia Team, Wikborg Rein Lawyers		X	Oslo	11.03.08	IO
						Murmansk	02.02.08	IO
51	Ane Koford Petersen	Senior Advisor Russia	Nordic Council of Ministers Copenhagen			Copenh.	12.03.08	NKM
52	Lemmi Tui	Senior Advisor Energy	Nordic Council of Ministers Copenhagen			Copenh.	13.03.08	NKM
53	Torgunn Oldeide	Vice President M&A	Statkraft	X		Oslo	26.03.08	NKM
54	Dag Sanne	General manager	Birkebeinerlaugets Bedriftsutvikling	X		Oslo	26.03.08	HK
55	Gudrun Knutsson	Energy Advisor	Swedish Energy Agency		X	Stockholm	27.03.08	NKM
56	Genady Shubin	Chairman of the Counsil of Directors	Kolaregionenergosbyt	X		Murmansk	01.04.08	IO
57	Alexey Presnov	Chairman of the Counsil of Directors	Kolaregionenergosbyt	X		Murmansk	01.04.08	IO
58	Tormod Briseid	Senior Researcher	Bioforsk	X		Murmansk	01.04.08	IO
59	John Brungot	Director	TideTec	X		Murmansk	01.04.08	IO
60	Erik Welle-Watne	Director	Innovation Norway	X		Murmansk	01.04.08	IO
61	Henrik Forsström	Senior Advisor	NEFCO	X		Murmansk	01.04.08	IO
62	Kari Homanen	Senior Investment Manager	NEFCO	X		Murmansk	01.04.08	IO
63	Svetlana Vorobyova	Senior Marketing Manager	Kolaregionenergosbyt			Murmansk	01.04.08	IO
64	Sergey Seliverstov	Legal Advisor	Sokolov, Maslov and Partners			Murmansk	01.04.08	IO
65	Paul Erik Aspholm	Head of Section for Biology	Bioforsk	X		Murmansk	01.04.08	IO
66	Pavel Ponkratyev	Deputy Head of Investment Project Implementation Unit	Hydro OGK			Murmansk	02.04.08	IO

Table A.6 Continued

#	Name	Position	Institution	Short informal conversation	Tel. interv.	Location	Date (d/m/y)	By
67	Vasily Shein	Leader of Engineering Work on Wind Energy	Mezen Small Tidal Power Station Inc.	X		Murmansk	02.04.08	IO
68	Ksenia Koloshtivina	Head of Programme and Investment Advisor	Nordic Environmental Finance Corporation			Murmansk	02.04.08	IO
69	Ksenia Boldyreva	Assistant	Murmansk State Technical University	X		Murmansk	02.04.08	IO
70	Natalia Trochinskaya	Assistant	Murmansk State Technical University	X		Murmansk	02.04.08	IO
71	Anastasia Shironina	Assistant	Murmansk State Technical University	X		Murmansk	02.04.08	IO
72	Lars Helge Helvig	Executive Manager	Norsk Vindenergi		X	Oslo	10.04.08	NKM
73	Bengt Hillring	Professor	Swedish University of Agricultural Science		X	Stockholm	13.04.08	NKM
74	Jelena Babajeva	Associate Banker	EBRD	X		Stockholm	15.04.08	NKM
75	Victor Balyberdin	Senior Analyst	Eclipse Energy Group	X		Stockholm	15.04.08	NKM
76	Andreas Thomas	General Manager	Vestas	X		Stockholm	15.04.08	NKM
77	Vladislav Norkin	General Director	ADD Research and Development efficient energy	X		Stockholm	15.04.08	NKM
78	Lars Landberg	R&D Director	Garrad Hassan	X		Stockholm	15.04.08	NKM
79	Jukka Nygren	Manager Finland	NORD POOL	X		Stockholm	15.04.08	NKM
80	Anatoli Kopylow	Ph.D.	RusHydro	X		Stockholm	15.04.08	NKM
81	G. Ermolenko	Ph.D.	Greta Energy Company	X		Stockholm	15.04.08	NKM
82	V. Belilovski	Ph.D.	Greta Inc.	X		Stockholm	15.04.08	NKM
83	A. Cherniovsky	Ph.D.	Rostov Teplo Electro Project	X		Stockholm	15.04.08	NKM
84	O. Popel	Ph.D.	Joint Institute for High Temperatures, Russian Academy of Science	X		Stockholm	15.04.08	NKM
85	Mikhail Zavorsky	Chairman of Directorial Board	JSC Nord Hydro	X		Stockholm	15.04.08	NKM
86	Johan Moss	Managing Director Eastern Europe	Tricorona		X	Moscow	16.04.08	NKM
87	Sergev Sheglov	Investment Manager	RaoNordic	X		Stockholm	16.04.08	NKM
88	Vladimir Turkin	General Director	ADD Research and Development efficient energy	X		Stockholm	16.04.08	NKM

Table A.6 Continued

Name	Position	Institution	Short informal conversation	Tel. interv.	Location	Date (d/m/y)	By
89 Hafsteinn Helgason	General Manager	Addison Engineering	X		Stockholm	16.04.08	NKM
90 Henrik Hjelm	Managing Director	Eastnet	X		Stockholm	16.04.08	NKM
91 Raili Kajaste	Special Advisor Energy and Environment	NEFCO		X	Helsinki	21.04.08	NKM
92 Ragnar Ottosen	Advisor	Rosnor	X		Oslo	25.04.08	NKM
93 Solveig Nordström	Vice President	Nordic Environmental Finance Corporation	X		Oslo	08.05.08	IO, NKM
94 Valery Kharchenko	Chief Scientific Officer	All-Russian Research Institute for the Electrification of Agriculture			Oslo	08.05.08	IO
95 Ottar Hermansen	Senior Advisor	Innovation Norway	X		Oslo	08.05.08	NKM
96 Erik Holthedal	Chairman of the Board	Scanteam	X		Oslo	08.05.08	HK
97 Maja Tofteng	Advisor	ECON Pöyry	X		Oslo	08.05.08	HK
98 Jørgen Thon	n.d.	Statnett	X		Oslo	08.05.08	HK
99 Harry Zilliacus	Senior Advisor	Nordforsk		X	Oslo	01.08.08	GD
100 Phil Saprykin	Deputy Director	Council of the Baltic Sea States		X	Stockholm	01.08.08	GD
101 John Christensen	Head of Centre	UNEP RISØ Centre on Energy, Climate and Sustainable Development		X	Roskilde	04.08.08	GD
102 Sigridur Thormodsdottir	Senior Advisor	Nordic Innovations Centre		X	Oslo	05.08.08	GD
103 Björn Gunnarsson	Academic Director	The School for Renewable Energy Science		X	Reykjavik	05.08.08	GD
104 Gudrun Knutsson	Senior Advisor	Swedish Energy Agency		X	Stockholm	03.08.08	GD
105 Bo Riisgaard Pedersen	Chief Programme Coordinator	Danish Energy Agency		X	Copenhagen	30.07.08	GD

Table A.7   Ten in-depth interviews, interviewees, sectors and organizations

Location	Organisation	Field	Type	Abbreviation
Russia	Nordic Council of Ministers office, Kaliningrad	Renewable energy	Governmental organization	A
Sweden	Swedish Energy Agency	Bioenergy	Governmental organization	A
Denmark	Nordic Council of Ministers office, Copenhagen	Energy/Russia	Governmental organization	A
Norway	Norwegian Wind Energy	Wind	Company	C
Norway	Statkraft	Hydropower	Company	C
Russia	TRICORONA	Carbon credit investments	Company	C
Norway	Rosnor	Hydropower/ advisors	Company	C
Finland	NEFCO	Environmentally friendly energy	Financial institution	F
Norway	Bellona	Renewable energy	Non-profit environmental organisation	NGO
Sweden	Swedish University of Agricultural Science	Bioenergy	University/ Research	U

- Where do you think the concepts of Nordic–Russian cooperation on renewable energy come from?
- How would you characterize Nordic–Russian cooperation within the field of renewable energy?
- Why Nordic cooperation in renewable energy with *Russia*?
- What *opportunities* exist?
- What are the *main reasons* for involving/ wanting to be involved in Nordic–Russian cooperation within renewable energy? (For your organization/for the companies)

3: Institutionalisation

- What institutions provide guidance for the cooperation your company/ institute/organization is involved in?
- Who initiated this collaboration?

4: Cooperation

- Why cooperate?
- Which type of Nordic–Russian cooperation are you involved in?
- How is the cooperation realized? (Organisational form, etc.)
- What specific areas are involved in your cooperation? (Resources, knowledge, technology, etc.)
- How does it work?
- Motivation/aim

- What do you believe is the motivation behind Nordic–Russian cooperation on renewable energy?
- What is the goal/objective of the cooperation?
- Partner selection *(If relevant)*
- How was the cooperative partner found?
- What determined/influenced the selection?

5: What are the challenges/barriers?

- What works, and what does not work, in the Nordic–Russian cooperation your organization is involved in?/for the companies?
- What are the main challenges in Nordic–Russian cooperation in the field of renewable energy (in general)?

6: Future of Nordic–Russian cooperation in RE

- What is the time horizon for your involvement in Nordic–Russian cooperation on renewable energy?
- Is there anything that could facilitate Nordic–Russian cooperation within the field of renewable energy?

7: Future of RE in general

- What are your views on renewable energy as a substitute for non-renewable sources

Bibliography

Aasland, Aadne. *Development in Research: An Outline of the Science Systems in Russia and the Baltic States.* Oslo: Nordforsk, 2007. http://www.nordforsk. org/_img/nordforsk_pb1_web.pdf [accessed 12 May 2008].

Abercade Consulting. *Rynok fotoelektricheskikh preobrazovateley i solnechnykh moduley Rossii 2005 Goda.* Moscow: Abercade Consulting, 2006.

Agence France-Presse. 'EU to Send Mission to Moscow and Kiev over Gas Dispute', 4 June 2009. http://news.id.msn.com/topstories/article.aspx?cp-doc umentid=3358147 [accessed 21 June 2009].

Agence France-Presse. 'EU–Russia Summit Fails to Mend Rifts', 27 May 2009. http://news.yahoo.com/s/afp/20090522/wl_afp/russiaeusummit_20090 522183530 [accessed 21 June 2009].

Amundsen, Eirik, Lars Bergman and Nils-Henrik M. von der Fehr. 'The Nordic Electricity Market: Robust by Design?', in *Electricity Market Reform: An International Perspective*, edited by Fereidoon Sioshansi and Wolfgang Pfaffenberger. Amsterdam: Elsevier, 2006, pp. 145–70.

Aron, Leon. 'Privatizing Russia's electricity'. *Russian Expert Review*, no. 4 (2003): 9–17.

Balyberdin, Victor. 'Russian Power Market: Structure, Development, Prospects'. Presentation held at the Energy Forum conference *Investing and Financing Renewable Energy in Russia*, Stockholm, Sweden, 15–16 April 2008.

Baranovskiy, Sergey and Aleksandr Chumakov. 'Alternative Energy in Russia: Problems and Perspectives'. *Alternativnaya Energetika*, vol. 7, no. 1 (2008): 2–6.

Barents Euro-Arctic Region. 'Information Document', 2005. http://www. barentsinfo.fi/beac/docs/11675_doc_CSO.2005.18ENGBarentsInfosustainabl edevlp.pdf [accessed 20 June 2009)], p. 2.

Barents Euro-Arctic Region. 'Cooperation in the Barents EuroArctic Region. Target: Stability and Sustainable Development', 1 June 2005. http://www. barentsinfo.fi/beac/docs/11675_doc_CSO.2005.18ENGBarentsInfosustainabl edevlp.pdf [accessed 20 June 2009].

BASREC. 'Baltic Sea Region Energy Co-operation', N.d. http://www.cbss.org/ Energy/baltic-sea-region-energy-cooperation [accessed 21 June 2009].

Bellona. 'Northwest Russia Renewable Energy Forum 1–2 April 2008'. http:// www.bellona.org/subjects/energy_forum [accessed 3 August 2008].

Bellona. Title Page. http://www.bellona.org/ [accessed 23 June 2009].

Bernard, Alain, Sergey Paltsev, John M. Reilly, Marc Vielle and Laurent Viguier. 'Russia's Role in the Kyoto Protocol', MIT Joint Program on the Science and Policy of Global Change, Report no. 98, June 2003.

Bezrukikh, Pavel. 'Netraditsionnye vozobnovlyaemye istochniki energii'. *Teplovoy Energeticheskiy Kompleks*, no. 4 (2001): 31–45.

Bond, Derek and Markku Tykkylainen. 'Northwestern Russia: A Case Study in "Pocket" Development'. *European Business Review*, vol. 96, no. 5 (1996): 54–60.

Borovaya, Zarema. 'Vystreli v tundre'. *Murmanskiy Vestnik*, 31 October 1998: 1–2.

British Council. *British Council Online Bulletin*, September 2007. http://64.233.183.104/search?q=cache:YEYdTBDJDO0J:www.ukro.ac.uk/insight/ei0709.doc+why+was+INTAS+discontinued&hl=en&ct=clnk&cd=7&client=firefox-a [accessed 30 May 2008].

Brown, Anna. 'Russian Renewable Energy Market: Design and Implementation of National Policy'. *Russian/CIS Energy and Mining Law Journal*, vol. 3, no. 6 (2005): 33–39.

BP. cited in *Russian Analytical Digest*, vol. 18, 2007.

Business Press. 'Solnechnaya Energetika'. *Energeticheskoe Prostranstvo*. N.d. http://www.energospace.ru/2008/05/22/solnechnaja-jenergetika.html [accessed 25 May 2008].

Business Press. 'Solnechnaya energetika rasshevelila rossiyskiy biznes'. *Novye Tekhnologii*, vol. 55, no. 428 (11 April 2008). http://businesspress.ru/newspaper/article_mId_37_aId_446548.html [accessed 25 May 2008].

Chirac, Jacques; Javier Solana, Romano Prodi and Vladimir Putin. 'Joint Declaration'. Brussels: EC, 10 October 2000. http://europa.eu/rapid/pressReleasesAction.do?reference=IP/00/1239&format=HTML&aged=0&language=EN&guiLanguage=en [accessed 21 June 2009].

Danilova, Svetlana. 'Akademicheskaya elita, Reyiting vuzov 2007'. N.d. http://www.ucheba.ru/vuz-rating/1922.html [accessed 23 June 2009].

Danish Energy Authority. http://www.ens.dk/da-DK/Sider/forside.aspx [accessed 21 June 2009].

Dannemand Andersen, Per and Birte Holst Jørgensen. *Grundnotat om metoder indenfor teknologisk fremsyn*. Risø: Forskningscenter Risø, 2001.

Dezhina, Irina. 'American Science Foundations in Russia as Driving Forces of International Transfer in Knowledge and Professional Skills'. Paper presented at the conference *Transforming Civil Society, Citizenship and Governance: The Third Sector in an Era of Global (Dis)Order*, University of Cape Town, South Africa, 7–10 July 2002. http://www.istr.org/conferences/capetown/volume/dezhina.pdf [accessed 5 March 2008].

EC. *The Lisbon Strategy*. Brussels: EC, 2000.

EC. *The Energy Charter Protocol on Energy Efficiency and Related Environmental Aspects: Regular review of Energy Efficiency Policies*. Brussels: EC, 2007a.

EC. *The European Research Area: New Perspectives*, Green Paper. Brussels: EC, 2007b.

EC. 'FP7 in Brief: How to Get Involved in the EU 7th Framework Programme for Research'. Brussels: EC, 2007c. http://www.ec.europa.eu/research/fp7/pdf/fp7-inbrief_en.pdf [accessed 21 June 2009].

EC. 'The Climate Action and Renewable Energy Package, Europe's Climate Change Opportunity', 8 January 2009. http://ec.europa.eu/environment/climat/climate_action.htm [accessed 21 June 2009].

Energy Information Administration. 'Russia Energy Data, Statistics and Analysis: Oil, Gas, Electricity, Coal'. http://www.eia.doe.gov/cabs/Russia/Background.html [accessed 9 June 2008].

Energy Information Administration. 'Russia Energy Profile', May 2008. http://tonto.eia.doe.gov/country/country_energy_data.cfm?fips=RS [accessed 9 June 2008].

EU. 'Issues Being Examined Under the Energy Dialogue', 2007. http://ec.europa.eu/energy/russia/overview/issues_en.htm [accessed 7 June 2008].

EU. 'Introducing the OPET Network'. N.d. http://cordis.europa.eu/opet [accessed 9 June 2008].

EU–Russia Technology Centre. *Renewable Energy Sources Potential in the Russian Federation and Available Technologies*. Moscow: EU–Russia Energy Technology Centre, 2004.

EU Work Programme. 'Cooperation Theme 5 Energy', 2008. ftp://ftp.cordis.europa.eu/pub/fp7/docs/wp/cooperation/energy/e_wp_200902_en.pdf [accessed 21 June 2009].

European Commission Delegation to Russia. *The EU and Russia: Exploring Beyond Borders*, Moscow, EC Delegation, 2006. http://www.delrus.ec.europa.eu/en/images/mText_pict/2/science%20eng.pdf [accessed 20 June 2009].

European Commission Delegation to Russia. 'Overview of Relations', 2007. http://www.delrus.ec.europa.eu/en/p_210.htm [accessed 3 June 2008].

Eurostat news release. 'Research and Development in the EU: Preliminary Results', 12 January 2007. http://europa.eu/rapid/pressReleasesAction.do?reference=STAT/07/6 [accessed 21 June 2009].

Gianella, Christian and William Tompson. *Stimulating Innovation in Russia: The Role of Institutions and Policies*. Working Paper. Paris: OECD, 2007.

Golub, Alexander and Elena Strukova. 'Russia and the GHG Market'. *Climatic Change*, no. 63 (2004): 223–43.

Graf Lambsdorff, Johann. *The Methodology of the Corruption Perceptions Index*. Berlin: Transparency International, 2007.

Hagen, Guro Aardal. 'Neppe et must for Google'. *Dagens IT* (2 June 2008). http://www.dagensit.no/finans/article1415158.ece?jgo=c1_re_left_6&WT.svl=article_title [accessed 6 June 2008].

Heleniak, Timothy. 'Migration and Restructuring in Post-Soviet Russia'. *Demokratizatsiya*, Fall (2001): 1–18.

Hemscott.com. 'Russian Domestic Gas Prices to Remain below European Export Level to 2014–15', 27 May 2008. http://www.hemscott.com/news/static/tfn/item.do?newsId=64497524037005 [accessed 7 June 2008].

Hill, Fiona and Clifford Gaddy. *The Siberian Curse: How Soviet Planners Left Russia Out in the Cold*. Washington, DC: Brookings Institution, 2003.

Hjelm, Henrik. 'Russian Business Culture: To Do Business in Russia'. Presentation at the conference *Investing and Financing Renewable Energy in Russia*, Stockholm, Sweden, 15–16 April 2008.

IEA. *Renewables in Russia: From Opportunity to Reality.* Paris: IEA, 2003. http://www.iea.org/textbase/nppdf/free/2000/renewrus_2003.pdf [accessed 29 February 2008].

IEA. *Russian Electricity Reform: Emerging Challenges and Opportunities.* Paris: IEA, 2005. http://www.iea.org/Textbase/publications/free_new_Desc. asp?PUBS_ID=1473 [accessed 29 February 2008].

Institute of Semiconductor Physics. Title Page, Siberian Branch, Russian Academy of Sciences. http://ecoclub.nsu.ru/altenergy/common/table2.htm [accessed 15 May 2008].

INTAS. *Report by External Evaluators on the Programme of the International Association for the Promotion of Co-operation with Scientists from the New Independent States of the Former Soviet Union (INTAS) in the Period 1993– 2003 to the INTAS General Assembly*, Brussels: INTAS, 1 October 2004.

International Institute for Applied Systems Analysis (IIASA) hompage. http:// www.iiasa.ac.at/docs/Research/ [accessed 8 August 2008].

Ioffe, Olga. 'Severny gorizont Rossii'. *Tekhsovet*, vol. 33, no. 2 (5 February 2006): 7–12.

Johannesen, Asbjørn, Line Kristoffersen and Per Arne Tufte. *Forskningsmetode for økonomiskadministrative fag.* Oslo: Abstrakt forlag, 2004.

Keikkala, Gudrun; Andrey Kask, Jan Dahl, Vladimir Malyshev and Viktor Kotomkin. 'Estimation of the Potential for Reduced Greenhouse Gas Emission in North-East *[sic]* Russia: A Comparison of Energy Use in Mining, Mineral Processing and Residential Heating in Kiruna and Kirovsk–Apatity'. *Energy Policy*, vol. 35, no. 3 (2007): 1452–63.

Khristenko, Victor and François Lamoureux. 'EU–Russia Energy Dialogue: Fifth Progress Report'. Brussels: EC, 2004. http://www.ec.europa.eu/energy/russia/ joint_progress/doc/progress5_en.pdf [accessed 8 June 2008].

Khutorova, Natalia. 'National Research Funding in Russia', Moscow State Forest University, 2008. http://www.ftpc5.si/files/FTP%20PDF%20Presentations/Tu esday%20parallel%20WoodWisdom/WW%20SLOVENIA%20RUSSIA.pdf [accessed 21 August 2008].

Kovalova, Natalia and Stanislav Zaichenko. 'The Russian System of Higher Education and its Position in the NSI'. Paper presented at *Universidad 2006: The 5th International Congress on Higher Education*, Havana, Cuba, 13–17 February 2006.

Kramer, Andrew E. 'Gazprom Moves into Coal as Way to Increase Gas Exports'. *New York Times*, 27 February 2008. http://www.nytimes.com/2008/02/27/ business/worldbusiness/27coal.html?ex=1361854800&en=17898f35b32aa09 b&ei=5088&partner=rssnyt&emc=rss [accessed 29 February 2008].

Kreutzmann, Anne. 'The Smell at the End of the World: Nitol Wants to Produce Silicon in Siberia'. *Photon International*, no. 11 (November 2007): 30–47.

Krivitsky, Sergey and Alexander Tsvetinsky. 'Oil and Gas Exploration on the Arctic Shore'. *Proceedings of the Eleventh International Offshore and Polar Engineering Conference, Stavanger: International Society of Offshore and Polar Engineers*, Stavanger, Norway, 17–22 June 2001, pp. 661–4.

Kroll, Luisa. 'The World's Billionaires'. *Forbes Magazine*, 5 March 2008. http://www.forbes.com/lists/2008/03/05/richest-people-billionaires-billionaires08-cx_lk_0305billie_land.html [accessed 25 June 2008].

Leal-Arcas, Rafael. 'EU Relations with China and Russia: How to Approach New Superpowers in Trade Matters'. *Journal of International Commercial Law and Technology*, vol. 4, no. 1 (2009): 22–52.

Littlechild, Stephen. 'Foreword: The Market Versus Regulation', in *Electricity Market Reform: An International Perspective*, edited by Fereidoon Sioshansi and Wolfgang Pfaffenberger. Amsterdam: Elsevier, 2006, pp. xvii–xxviii.

Lo, Bobo. 'Evolution or Regression? Russian Foreign Policy in Putin's Second Term', in *Towards a Post-Putin Russia*, edited by Helge Blakkisrud. Oslo: NUPI, 2006, pp. 57–77.

Luo, Yadong. 'Building Trust in Cross-cultural Collaborations: Toward a Contingency Perspective'. *Journal of Management*, vol. 28, no. 5 (2002): 669–94.

Makarov, Alexey. *New Energy Consumption and Supply Trends (Worldwide and Russia)*. Moscow: Energy Research Institute, Russian Academy of Sciences, 2004.

Marchenko, O.V. and S.V. Solomin. 'Efficiency of Wind Energy Utilization for Electricity and Heat Supply in Northern Regions of Russia'. *Renewable Energy*, vol. 29, no. 11 (2004): 1793–809.

Martinot, Eric. 'Energy Efficiency and Renewable Energy in Russia: Transaction Barriers, Market Intermediation, and Capacity Building'. *Energy Policy*, vol. 26, no. 11 (1998): 905–15.

Massari, Maurizio. 'Russia and the EU Ten Years On: A Relationship in Search of Definition'. *The International Spectator*, vol. 42, no. 1 (March 2007): 1–15. http://www.iai.it/pdf/articles/massari.pdf [accessed 21 June 2009].

Mckeown, Max. *The Truth about Innovation*. Harlow: Pearson Education Limited, 2008.

Medetsky, Anatoly. 'Gazprom to Delay Field Due to Low Demand'. *Moscow Times*, 16 June 2009. http://www.themoscowtimes.com/article/1009/42/378835.html [accessed 21 June 2009].

Medvedev, Alexander I. 'Growth Solutions for the Low Carbon-Economy'. Keynote speech, *Global Leadership and Technology Exchange* meeting, Moscow, Russia, 4 July 2008.

Merkina, Natalia. 'Innovation and Regional Development in Russia'. MSc thesis, University of Oslo, Environmental and Development Economics, 2004.

Merle-Béral, Elena. 'Russia Renewable Energy Markets and Policies: Key Trends'. Presentation at *Global Best Practice in Renewable Energy Policy Making*, expert meeting, Paris, France, 29 June 2007. http://www.iea.org/Textbase/work/2007/ bestpractice/Merle_Beral.pdf [accessed 23 July 2008].

Merle-Béral, Elena. 'The Wider Perspective: Russia's Energy Scene'. Presentation at the conference *Renewable Energy in Russia: How Can Nordic and Russian Actors Work Together?*, Oslo, Norway, 8 May 2008.

Ministry of Energy of the Russian Federation. *Summary of the Energy Strategy of Russia for the Period of up to 2020.* Moscow: Ministry of Energy of the Russian Federation, 2003.

Moe, Arild. 'The Kyoto Mechanisms and Russian Climate Politics'. Presentation at the conference *Renewable Energy in Russia: How Can Nordic and Russian Actors Work Together?*, Oslo, Norway, 8 May 2008.

Nansen Environmental and Remote Sensing Centre homepage. http://www.nersc. no/main/index2.php [accessed 28 July 2008].

Nansen International Environmental and Remote Sensing Centre. 'About NERSC'. http://www.nersc.no/main/index2.php?display=aboutsummary [accessed 28 July 2008].

Nansen International Environmental and Remote Sensing Centre. 'Scientific Foundation'. http://www.niersc.spb.ru/niersc/index.php?section=main [accessed 30 July 2008].

Nikitin, Oleg. 'Kak perevozitsya neft'. *Tekhsovet*, vol. 33, no. 2 (5 February 2006): 2–4.

Nordic Environment Finance Corporation (NEFCO). 'Carbon Finance'. http:// www.nefco.org/financing/carbon_finance [accessed 31 July 2008].

Nordic Environment Finance Corporation (NEFCO). 'Financing'. http://www. nefco.org/financing [accessed 31 July 2008].

Nordic Environment Finance Corporation (NEFCO). 'Funding Resources'. http:// www.nefco.org/introduction/funding_resources [accessed 25 July 2008].

Nordic Environment Finance Corporation (NEFCO). 'Introduction'. http://www. nefco.org/introduction [accessed 31 July 2008].

Nordic Environment Finance Corporation (NEFCO). 'Nordic Environmental Development Fund'. http://www.nefco.org/financing/NMF [accessed 31 July 2008].

Nordic Energy Research. *Strategic Action Plan 2007–2010.* Oslo: Nordic Energy Research, 2006.

Nordic Energy Research. *Annual Report 2007.* Oslo: Nordic Energy Research, 2008.

Nordic Investment Bank homepage. http://www.nib.int/home [accessed 23 June 2009].

Nordic Investment Bank. 'Lending'. http://www.nib.int/lending/neighbouring. html [accessed 3 August 2008].

Nordic Investment Bank. 'News'. http://www.nib.int/newsen/1211961178.html [accessed 3 August 2008].

Northern Dimension Environmental Partnership (NDEP). 'Background'. http:// www.ndep.org/home.asp?type=nh&pageid=5#how2 [accessed 30 July 2008].

Northern Dimension Environmental Partnership (NDEP). 'Project Pipeline'. http://www.ndep.org/projects.asp?type=nh&cont=prjh&pageid=15&content= projectlist [accessed 20 July 2008].

Northern Dimension Environmental Partnership (NDEP). 'The Structure of NDEP'. http://www.ndep.org/home.asp?type=nh&pageid=6 [accessed 30 July 2008].

Novye energeticheskie proekty. 'Natsionalnaya innovatsionnaya programma'. http://www.nic-nep.ru/ [accessed 23 June 2009].

Ó Tuathail, Gearóid. 'Russia's Kosovo: A Critical Geopolitics of the August 2008 War over South Ossetia'. *Eurasian Geography and Economics*, vol. 49, no. 6 (2008): 670–705.

Oldfield, Jonathan D. *Russian Nature: Exploring the Environmental Consequences of Societal Change.* Aldershot: Ashgate, 2005.

Oresheta, Mikhail. 'Umchi menya, olen'. *Polyarnaya Pravda* (28 November 1996), pp. 1–2.

Øverland, Indra. 'Russia's Hydrogen Sector'. Presentation at the conference *Renewable Energy in Russia: How Can Nordic and Russian Actors Work Together?*, Oslo, Norway, 8 May 2008. http://english.nupi.no/content/ download/4431/61693/file/Indra%20Overland%20[Read-Only].pdf [accessed 22 June 2009].

Perović, Jeronim and Robert Orttung. 'Russia's Energy Policy: Should Europe Worry?'. *Russian Analytical Digest*, no. 18 (2007): 2–7.

Piper, Jeff. 'Toward an EU–Russia Energy Partnership'. Presentation at the conference *Energy Security: The Role of Russian Gas Companies*, Paris, France, 25 November 2003.

Pittman, Russell. 'Restructuring the Russian Electricity Sector: Re-creating California?'. *Energy Policy*, vol. 35, no. 3 (2007): 1872–83.

Popel, Oleg S. 'Tekhnologii i sfery effektivnogo energeticheskogo ispolzovaniya vozobnovlyaemykh istochnikov energii v regionakh Rossii'. Presentation at the seminar *Itogi realizatsii proektov v ramkakh prioritetnogo napravleniya 'Energetika i energosberezhenie' federalnoy tselevoy nauchno-tekhnicheskoy programmy v 2007 g.*, Moscow, Russia, 7 December 2007.

Potanin, Vladimir. 'Fond Potanina sostavil reyting vedushchikh rossiyskikh vuzov'. N.d. http://www.ucheba.ru/vuz-rating/1922.html [accessed 23 June 2009].

RAO UES. (United Energy Systems of Russia). *Concept of RAO UES Strategy for 2003–2008: '5+5'*, Moscow: RAO UES, 2003. http://www.rao-ees.ru [accessed 29 February 2008].

RAO UES. (United Energy Systems of Russia) webpage. http://www.rao-ees.ru [accessed 15 May 2008].

Ripinsky, Sergey. 'The System of Gas Dual Pricing in Russia: Compatibility with WTO Rules'. *World Trade Review*, vol. 3, no. 3 (November 2004): 463–81.

Rosbalt News Agency. *Mintrans Raportuet o Vypolnenii Severnogo Zavoza.* Press release, 28 September 2007. http://www.rosbaltnord.ru/2007/09/28/417963 [accessed 4 March 2008].

Runge, Tatiana. *ININ Results, Achievements and Recommendations*. Brussels, ININ, 28 November 2006. http://www.intas.be/documents/ininworkshops/6inin_ workshop_presentations/ININ_results_achievements_&_recommendations. pdf [accessed 20 May 2008].

Russian Courier. 'Putin's Aid: Kyoto Protocol is Totalitarian', 19 May 2004. http://www.gateway2russia.com/st/art_236288.php [accessed 5 May 2008].

Russian Courier. 'Russia Launches Advanced Energy Market', 1 September 2006. http://www.russiancourier.com/en/news/2006/09/01/59466/ [accessed 5 May 2008].

Russian Analytical Digest. 'Russia's Energy Policy', no. 18 (3 April 2007): 1–17.

Semiconductor Equipment and Materials International (SEMI). *Russia Market Update*, 26 February 2008. http://content.semi.org/cms/groups/public/ documents/events/p042796.pdf [accessed 25 May 2008].

Shuster, Simon. 'Gazprom, Interros Ready to Carve Up Power Industry'. *St Petersburg Times*, 13 February 2007. http://www.sptimes.ru/index.php?action_ id=2&story_id=20354 [accessed 21 June 2009].

Sioshansi, Fereidoon P. 'Electricity Market Reform: What Have We Learned? What Have We Gained?'. *The Electricity Journal*, vol. 19, no. 9 (2006): 70–83.

Sioshansi, Fereidoon P. and Wolfgang Pfaffenberger. *Electricity Market Reform: An International Perspective*. Amsterdam: Elsevier, 2006.

Skurbaty, Tim. *Understanding the EU–Russia Energy Relations: Conflictual Issues of the ED and the ECT.* MA thesis, University of Lund, Department of Political Science, 2007. http://theses.lub.lu.se/archive/2007/05/16/1179319221-4953-154/MEA_thesis.pdf [accessed 7 June 2008].

Smith, Howard. *What Innovation Is: How Companies Develop Operating Systems for Innovation*. CSC White Paper for the European Office of Technology and Innovation, Brussels, 2007.

Starkov, Aleksandr, Lars Landberg, Pavel Bezroukikh and Mikhail Borisenko. *Russian Wind Atlas*. Roskilde: Risø National Laboratory, 2000.

Stern, Jonathan. 'The Russian–Ukrainian Gas Dispute', Oxford Institute for Energy Studies, 16 January 2006. http://www.oxfordenergy.org/pdfs/comment_0106. pdf [accessed 22 June 2009].

Strebkov, Dmitry S. 'Large-Scale Renewable Energy Technologies'. Presentation at the *3rd International Conference on Materials Science and Condensed Matter Physics*, Chisinau, Moldova, 3–6 October 2006.

Swahn, Natalia. *The Role of Cultural Differences between Norway and Russia in Business Relationships: Application to Strategic Management in Norwegian Companies*. Ph.D. dissertation, Norwegian University of Science and Technology, Trondheim, Faculty of Social Sciences and Technology Management, 2002.

Swedish Agency for Public Management. *Baltic Billion Fund 2: A Final Assessment.* Stockholm, 2006. http://www.statskontoret.se/upload/ Publikationer/2006/200608_englishsummary.pdf [accessed 21 June 2009].

Takla, Einar. 'Nå har vi nok'. *Dagens Næringsliv*, 25 August 2008. http://www.dn.no/forsiden/borsMarked/article1475178.ece?WT.mc_id=dn_rss [accessed 29 August, 2008].

Tompson, William. *Restructuring Russia's Electricity Sector: Towards Effective Competition or Faux Liberalisation?* OECD Economics Department Working Papers No. 403, 2004.

Ucheba.ru. 'Reyting vuzov po kriteriyu usloviya obucheniya v vuze'. N.d. http://www.ucheba.ru/vuz-rating/2856.html [accessed 23 June 2009].

UN Framework Convention on Climate Change website. http://unfccc.int/kyoto_protocol/mechanisms/items/1673.php [accessed 25 August 2008].

UN Statistics Division. http://unstats.un.org/unsd/cdb/cdb_series_xrxx.asp?series_code=30248 [accessed 9 June 2008].

UNEP/RISØ. 'JI Projects'. http://cdmpipeline.org/ji-projects.htm [accessed 14 August 2008].

UNESCO. *UNESCO Science Report*, Paris, 2005.

UNESCO Institute for Statistics. http://stats.uis.unesco.org/unesco/tableviewer/document.aspx?FileId=76 [accessed 8 June 2008].

Vetlesen, Arne Johan. 'Nullsumlogikk og andre narrespill–norsk oljepolitikk i klimakrisens tjeneste'. *Samtiden*, no. 2, 2008: 54–64.

Watkins, Alfred. *From Knowledge to Wealth: Transforming Russian Science and Technology for a Modern Knowledge Economy*. World Bank Policy Research Working Paper no. 2974, 2003.

Wengle, Susanne. 'Power Politics: Electricity Sector Reforms in Post-Soviet Russia', *Russian Analytical Digest*, no. 27 (2007): 6–9.

World Bank. *World Development Report 2008: Agriculture for Development*. Washington, DC: World Bank, 2008.

World Bank in Russia. *Energy Efficiency in Russia: Untapped Reserves*. Moscow: World Bank, 2008.

Xu, Yi-Chong. *Electricity Reform in China, India and Russia: The World Bank Template and the Politics of Power*. Northampton, MA: Edward Elgar, 2004.

Index

Milton Keynes UK
Ingram Content Group UK Ltd.
UKHW031132141024
449569UK00006B/238